THE

NICOLSON PAVEMENT,

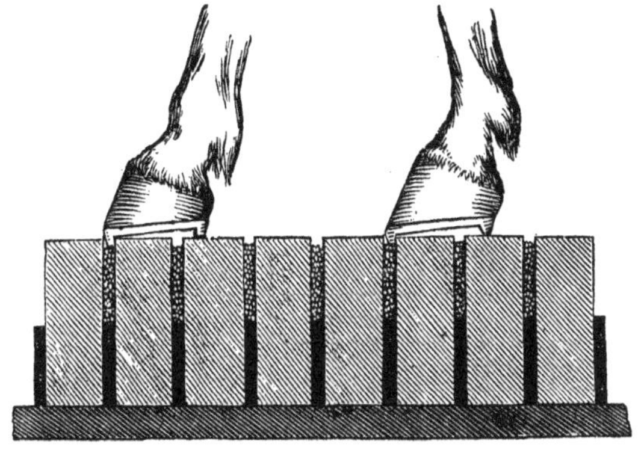

AND

PAVEMENTS GENERALLY,

BY

FRANK G. JOHNSON, M. D.

Any pavement that greatly increases the destruction of shoe, horse, vehicle, ease, comfort and convenience is not economical, though it cost nothing and last forever.

NEW YORK:
W. C. ROGERS & CO., PRINTERS AND STATIONERS,
No. 26 John Street,
1867.

PREFACE.

Believing there is need of a radical change in the method of paving, and, since the commencement of its introduction in New York city, a few months since, the Nicolson Pavement having given universal satisfaction, and elicited from the people innumerable expressions of gladness at the prospect of having better streets; and many inquiries having been made as to its cost, merits, durability, &c., by thousands who do not, as yet, comprehend its true character; and many persons, though *admiring* it, still doubting its durability, for the mere reason that it is essentially composed of wood; and some going so far, for the same reason, as to condemn it altogether, even before seeing it at all; and, for the same reason, freely admitting ourselves, that its *durability*, as well as some of its other merits, is not, at first, so self-evident or easily appreciated; and yet knowing, as we do, that its durability, and other superior qualities, have been proven by years of actual test, as well as being susceptible of scientific demonstration, we have prepared the following pages, to set forth a few principles relating to pavements generally; but principally to describe the construction, explain the advantages and demonstrate the character of the *Nicolson Pavement.*

Though the word *pavement*, (from *pavio*, signifying to beat or ram down), includes more than the streets of cities, taking in roads extending hundreds of miles, park roads, foot walks, &c.; yet, this pamphlet will be confined principally to that branch of the subject embracing *city* pavements.

While we might have said less, there is one subject connected with pavements to which we have barely alluded, but on which much might and ought to be written. We refer to the cruelty and inhumanity of driving horses on stone pavements.

NEW YORK, February 27th, 1867.

CONTENTS.

INTRODUCTION.

REQUISITES OF A GOOD PAVEMENT.

First Requisite of a Good Pavement is a Flat and Even Surface.

Second Requisite of a Good Pavement is, to have it so provided with Grooves or Cells as to afford a sure Foot-hold for the Horse.

Third Requisite of a Good Pavement is to have it slightly Arched, and provided with suitable Gutters.

Fourth Requisite of a Good Pavement is Stability of Structure: that is, its several Parts must be so formed and combined together that, when put down, they will retain their Original Position.

DESCRIPTION OF THE NICOLSON PAVEMENT.

INTRODUCTION.

Importance of Good Pavements.

The streets of a city are public property. Every citizen owns and makes daily use of them. They are the only parts of a city in which all are practically interested. In point of *necessity* no other public work can equal them. Yet the advantages of good pavements are overlooked and neglected.

The amount of travel on any railroad is many times less than that on any principal streetof a city; for, over the road, there goes only *now and then* a train, while the street is *ever crowded* with men, women, and children, horses, and vehicles, constantly moving to and fro; yet compare the enormous expense of three or four *hundred* miles of railroad with the paltry cost of *only* three or four miles of pavement. The traffic and travel on all the gorgeous steamers dashing up and down the waters of the Hudson, cannot equal the freight and passengers borne through a single thoroughfare of a city, on wheel and axle. Yet the expense of paving a street is but a trifling part of the cost of these steamers, to convert the river to a highway.

The streets of a city, taken with edifices fronting on them, suggest the idea of magnificent canals, deep but narrow; and on the side walls of which have been expended millions of dollars for substantiality and ornamentation; while, for perfecting the bed or bottom of which, scarcely one per cent. of an equal expense has been applied.

Through these canals there is a ceaseless rush of the tide of life. Year after year, surging masses of beings and objects, man and beast and vehicle, are promiscuously slipping, scrabbling, and jolting, rattling and thundering along, with nought whereon to tread and roll but endless piles of stones.

Between the side walls of any principal street of a large city, during the existence of these structures there pass and repass people, animals, and vehicles, equal in the aggregate, to all such in the world; yet never does one of these so much as *touch* upon the fronts of these buildings, while every foot and wheel must tread and roll, step by step and inch by inch, over the imperfect, rough, gullied and stony track beneath.

Hence, the most indispensable feature of public streets is that of Pavements. The convenience and comfort of the public can, therefore, be in no other way so much promoted as by radically improving them. For all will admit that the art of paving is behind the times; there being nothing, of so much importance, so poorly done.

The present mode of paving is violative of philosophic and mechanical principles, and miserably defective; there being scarcely a paved street in any city through which it is a pleasure to ride; or, in consequence of noise, dust and mud, to walk.

This is not in keeping with the spirit of the age; for, in the building of modern cities, skill and expense have been freely lavished in perfecting every public convenience, save that of paving streets. In fact, man, with ingenuity and art, has come to apply the inanimate forces to such an extent, that his strength is as illimitable as the boundless power of steam. Versed in science, and equipped with iron charged with the elements of nature, he tunnels rivers, and spans them with granite arches, big as the rainbow; he can move hills, and excavate the earth beneath its waters; can keep the sea *out*, or shut it *in;* can stretch an inclined water-track from lakes to ocean, whereon to slide his ponderous traffic; can drive his floating palaces from shore to shore; can find his way among mountains of ice to the poles; can bore the solid mountains as with a gimblet, and so walk *through* or *over* them, as suits his convenience; can besiege a city and destroy it in a night, and *make* a city in a fortnight; can pour a flood of water upon a burning mass, and thus make the elements overpower each other; can lay under ground arteries of iron through a vast city, and fill them with burning gas, that lights every dark hole, street and palace; can divert and bring down to the metropolis whole rivers, and drive them a thousand feet high, and supply every house with living streams of water; can stretch his telegraphic wires and iron track over valleys and mountains, from ocean to ocean; can construct giant printing presses, and, with the speed of lightning, ink his ideas and scatter them from brain to brain; can run his electric thread of discourse with other nations beneath the waters of the Atlantic; and can even aspire to leap the ocean and the desert with flying engines—yet he cannot, or at any rate, he *has* not, invented a pavement for the thronged streets of his beautiful and magnificent cities, that is not at once destructive of ease and comfort, a violation of the first principles of economy, an outrage to the laws of humanity, and a disgrace to an age of civilization.

Bearing in mind the aggregate qualities and requisites of a good pavement, the ingenuity of man cannot invent another so *poorly* adapted, to the wants of modern cities, as the *cobble* stone pavement; notwithstanding it is the kind principally employed.

Considering the present development of the arts and sciences, there is no reason why city streets should not be as agreeable, for *walking* and *pleasure riding*, as the roads in Central Park; and, at the same time, be so *substantially* paved as to meet all other necessary requirements.

A pavement to comport with other city improvements and conveniencies, such as gas, public water, &c., should be, practically, equivalent to a universal rail track.

What principally determines the nature of Pavements.

Though there be many requisites of a good pavement, yet the more essential of these are determined by the unchangeable nature of the horse and the shape of his foot, together with the indispensable circular form of the wagon wheel; and the fact, that both *foot* and *wheel* are *unavoidably* shod and bound with *iron*. Hence, any pavement, whatever be the mode of its construction and nature of its materials, to be *philosophically* adapted, must supply, at least, the following

Indispensable qualities.

It must afford and maintain an even and flat surface for foot and wheel, and yet, provide a sure *foot-hold* for horses. It must be firm and hard, to resist the action of iron shoes and tires; and yet, it must not possess that *rock-like* solidity which renders the resistance too shocking to the horse, and the noise of the horse and vehicle too deafening to man.

These, and other considerations of less importance, such as cost, cleanliness, durability, &c., will be discussed somewhat in detail, as we pass on.

Pavements of other Times.

Some historians hold that paved roads were first made by the Egyptians ; others, by the Carthaginians, but this is not material.

About two hundred years after the expulsion of their kings, the Romans introduced pavements ; and, for durability and evenness of surface, constructed the best paved roads ever known, before or since.

The most ancient and celebrated of which was the Appian way, called the "Queen of Roads." This was constructed in the year 311 before the birth of Christ. It extended from Rome to Capua, thence to Brundusium, a distance of three hundred and fifty miles, and was from eight to fifteen feet wide.

These pavements consisted of three distinct layers of materials ; the lowest was stone mixed with cement, and nine inches thick ; the middle, gravel or small stones and cement, six inches thick, to prepare a level and unyielding surface to receive the upper and most important structure, which consisted of large polygonal blocks, accurately fitted together.

The good condition of Roman roads was considered of so much importance that the charge was only intrusted to persons of the highest dignity ; and Augustus himself assumed the care of those in the neighborhood of Rome. "The expense of their construction was enormous." What a reflection is this on the management and condition of the streets of modern cities—especially New York!

"Though the paved roads of the ancient Romans surpass all other structures of the kind that have been made by civilized nations since their time, yet there are found in Peru remains of works of a similar construction of unknown age, and exceeding them in grandeur and extent. Such were the great roads from Quito to Cuzco, and continued south towards Chili, laid out through mountainous and almost impassable regions for distance variously estimated from *fifteen hundred to two thousand miles.*" They were built of heavy flags of freestone.

In Central America, among the ruins of Palenque, are also found pavements of large square blocks of stone constructed with great skill and nicety. In the middle ages but little attention was given to the paving of streets in Europe. Cordova in Spain was paved in 850. Streets in Paris were first paved in 1184. Many of the streets in London were in a perilous condition, by reason of deep pits and sloughs, even as late as the sixteenth and seventeenth centuries. Holborn was first paved in 1417; but the great market of Smithfield remained without pavement two hundred years longer.

The various methods of paving, and the different materials employed, will be described and their merits considered in discussing the requisites of a good pavement.

REQUISITES OF A GOOD PAVEMENT.

First Requisite of a Good Pavement is a Flat and Even Surface.

There are several reasons why such a surface is necessary:—

1st Reason why a flat and even surface is necessary is, that the Form and Action of Wagon Wheels require it.

In forming wagon wheels the aim is to produce a geometrical circle; that is, a periphery every point of which shall be equally distant from the center of motion, that the axle of the vehicles may be carried in a horizontal plane; which is easily accomplished, provided the periphery passes over a perfect plane also.

But if the rim of the wheel passes over an imperfect plane the circular form of the wheel is destroyed.

It matters but little whether the wheel be *perfect* and the surface of the pavement *imperfect*, or the surface of the *pavement* perfect and the *wheel* imperfect, so long as the perfection of either is indispensable to the perfecting of the other.

"*Humpyty-bump*" is no more desirable than "*bumpyty-hump*."

In this view of the subject, a cobble stone pavement, though it constitutes the great majority of all modern pavements, is the very poorest that can possibly be made, if not invented.

Cobble paving stones average about nine inches in

diameter, and, being nearly round in form, will present but a single point of contact with the wheels of wagons or the runners of sleighs; even though they be laid so accurately as to bring the crown of every stone into one common plane—which, by the way, it is not possible to do, and if it were they would not remain so.

Hence, a wheel comes into contact with such a pavement *not* in a straight *line* of a plane, but on *points* of a plane, nine inches asunder; which, in effect, destroys the circular character of the wheel and converts it into a polygon, with sides nine inches long. Suppose a wheel to be four feet in diameter, this polygon would have sixteen sides. Now, suppose the wheel to have sixteen spokes, these sides will reach from spoke to spoke; therefore, a *perfect* wheel on such a pavement, is no better, practically, than a *sixteen-sided polygon* would be on a pavement of a *flat and even* surface: as will be illustrated by the following diagram; which is drawn on a scale of one inch to the foot.

AA represents the rim of the wheel, which is 4 feet in diameter; DD, the paving-stones, which are 9 inches in diameter; with the crown of each lying in the straight dotted line CC.

If the pavement were flat, every point of the periphery of the wheel would come into contact with this straight line; but in passing over the cobble stones, DD, it sinks down between the crowns of the stones; as shown by the *curves* in the line CC; which will cause the center of the axle to *rise* and *fall*, so as to describe the *zigzag* line BB.

Now, as the distance of one spoke from another, (nine inches) is the same as it is from crown to crown

Fig. 1

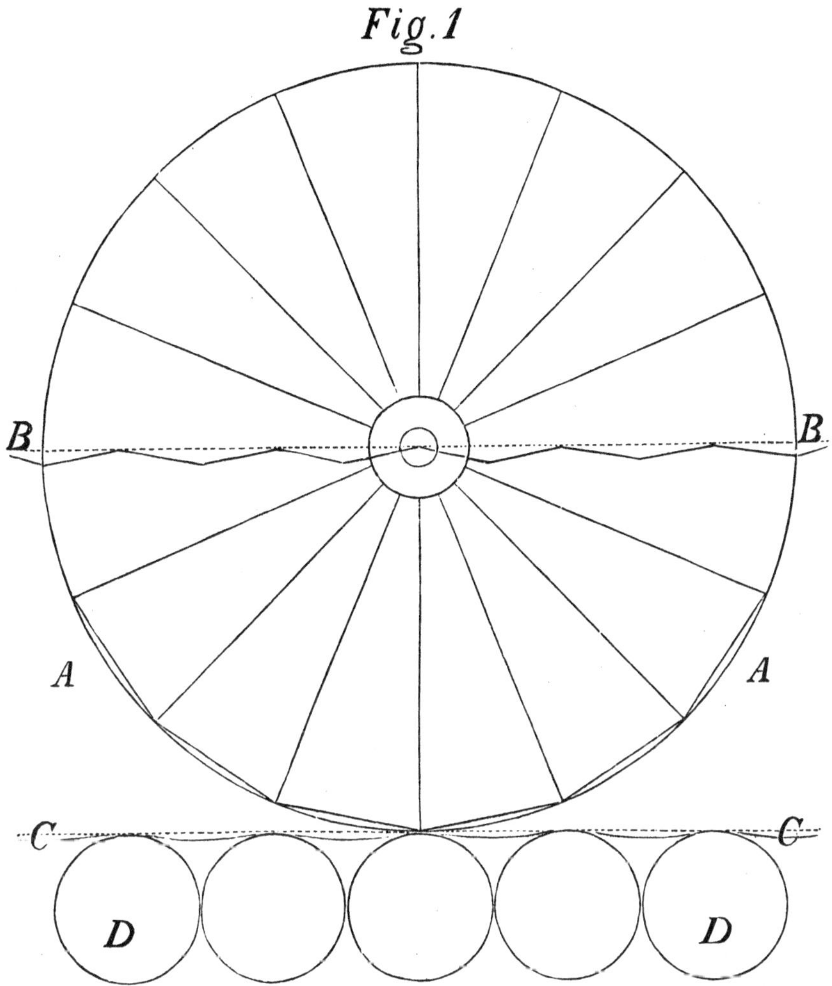

of the stones, it would make no difference, with the motion of the *axle*, if the rim of the wheel were taken off; provided the ends of the spokes were made to strike on the crowns of the stones; which would be the same as converting the wheel, (by drawing straight lines from spoke to spoke) into a polygon, of sixteen nine-inch sides, and then rolling it over a *flat* pavement.

Such a pavement is wholly unmechanical, and is but little better than a country road, filled with small logs, to keep wheels out of the dirt and mud; and riding through a city, thus paved, is like driving over an endless stone heap.

As regards its relation to *wheels*, the Belgian pavement which, in amount laid, comes next to the cobble stone, is not much better, unless it be laid with great skill and care. For, although its general surface is more even than that of the other, yet the following illustration will show that many of the pieces, or blocks, being unevenly cut, have only a single point of contact with the wheel; for which reason, (so far as it applies,) it has the same effect as the cobble stone pavement in throwing the motion of the axle out of a straight line; and which is illustrated by the following cut, drawn on the same scale as the other.

It will be seen that each of the blocks, has one or more points higher than the rest of its surface; shown by the letters r, s, t, u, v. Between these points are corresponding depressions, shown by the figures 1, 2, 3, 4, 5. Most of these depressions are sufficiently deep to prevent the curvature of the wheel from reaching to the bottom: hence, on an average, the wheel comes into contact with the pavement only on *one* point to a stone.

It matters not how slight may be the depression between any two of these points, so long as the wheel does not touch the surface *between* the points. Therefore the motion of the axle will be thrown out of a straight line, as shown by the line BB; but not so

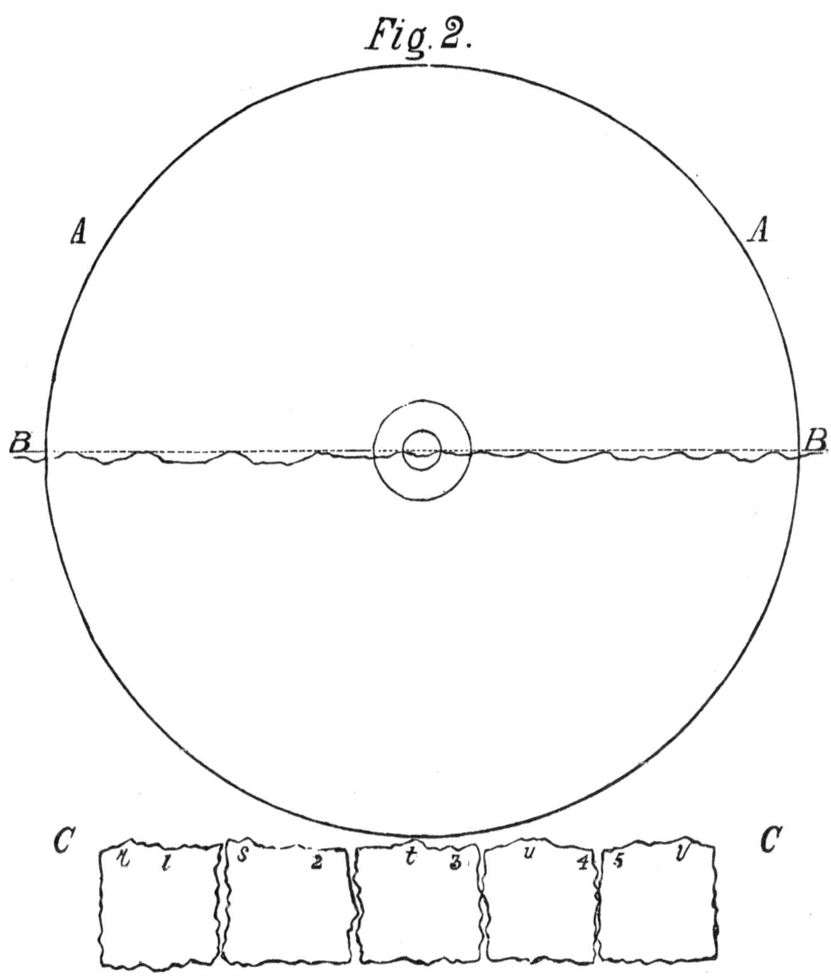

Fig. 2.

much as on the cobble pavement; for the reasons that, the oval form of the cobble stone brings the highest point of each stone near its verticle axis, which prevents them from coming nearer together than by a distance equal to the diameter of the stones; while many of the stones settle below those around them, and this increases the space between the points of contact.

But with the Belgian pavement, the blocks, being square and smaller, bring the points of contact somewhat closer together; and, being irregularly uneven, destroy the uniformity of the spaces between the points of contact: and, again, the pieces being square, or nearly so, admit of their being laid so as to remain somewhat longer in a uniform position.

Nevertheless; this style of pavement, as it affects the action of the wheel and axle, is not a radical remedy of the defects of the cobble stone pavement, but only an improvement, a step in advance of it; and, in the same manner and for the same cause, keeps the axle going constantly *up and down*, instead of carrying it, as it should, and as it is possible to be done, along in a perfect horizontal plane.

By the following illustration (Fig. 3), which represents a longitudinal section of the Nicolson pavement, it is shown that the axle of vehicles can be borne along in a horizontal plane.

DD represent the blocks or pieces of the pavement, showing that their upper surfaces, (as well as all their other faces,) are perfectly flat and lie in one common horizontal plane. The cells or grooves, GG, which furnish foothold for the horse, being only three-quarters of an inch wide, the curvature of the wheel does not sink between the blocks *at all;* and, consequently the axle, instead of describing a *zigzag* line, as it does on the cobble and Belgian pavements, passes along the straight line BB of a plane, parallel to the surface of the pavement.

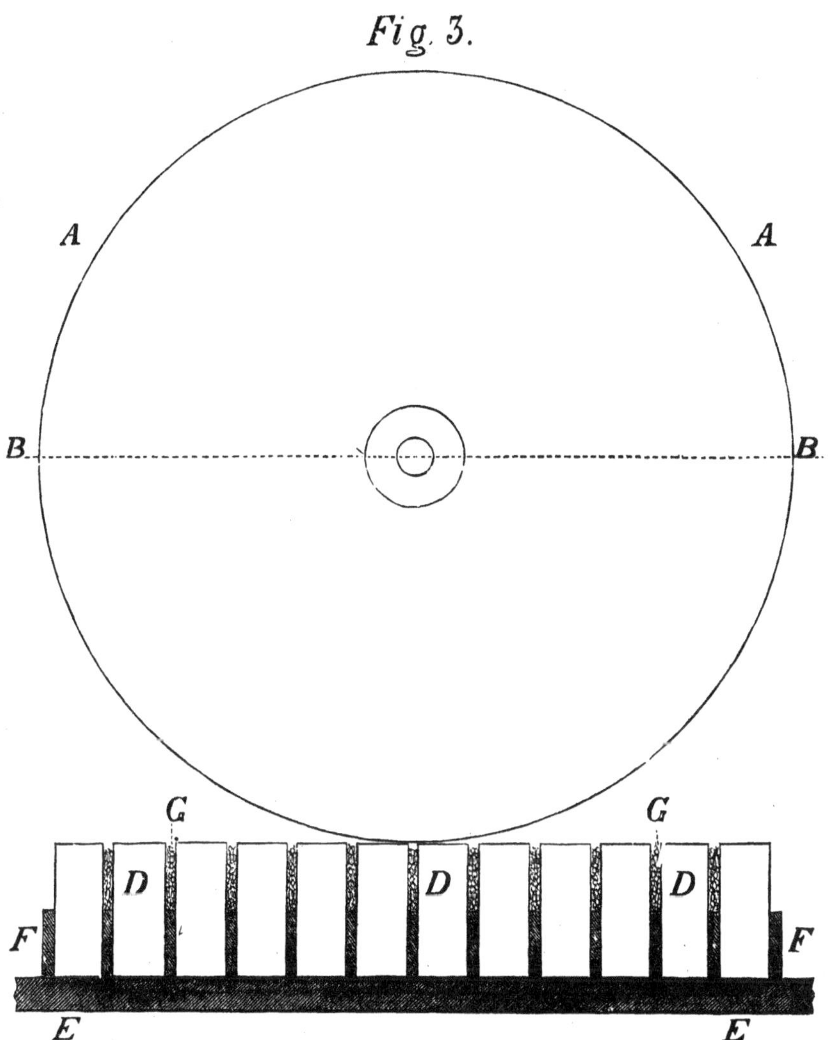

Fig. 4 represents a transverse section of the Nicolson; showing, again, that its surface is perfectly flat and level.

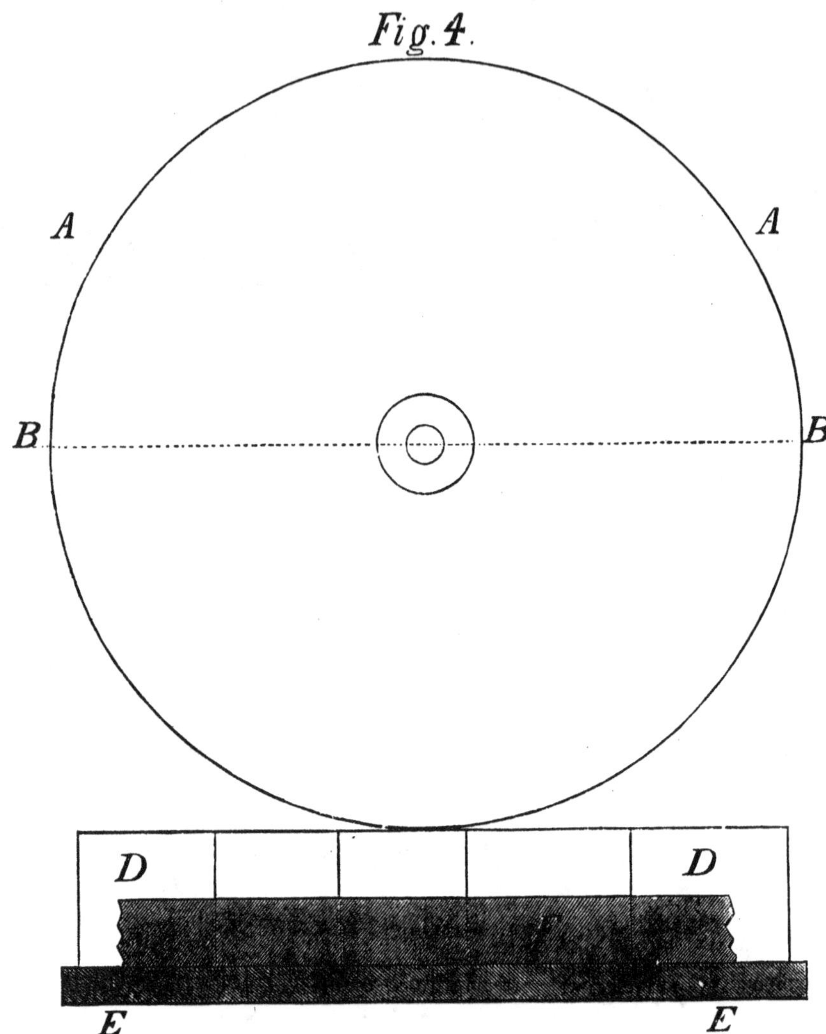

2d Reason why a flat and even surface is necessary is, to diminish the Obstruction to Vehicles, and so render the Draft of Horses more Effective.

On a perfectly level and flat pavement one horse is

as powerful as two or three on an uneven surface. On the cobble stone pavement, if it is in good order, the vehicle, by being kept in rapid motion, will roll from stone to stone with about the same ease that it would on a flat surface. But the *starting* of a loaded wagon, when the wheels stand between the stones, or are down in some gully or depression, is quite another matter; and it requires often many times the force to start the load that it would were it resting on a plane surface.

By referring to Fig. 1 (page 19) it will be seen that the cobble stone pavement, in this respect also, is the *most* objectionable; for each wheel of a vehicle, on being stopped, will rest between two stones, in such a manner that they may be said to be blocked or "chocked." Hence before a vehicle can be started it must first be *lifted*, so that it often requires the force of several horses to *start* a load that one horse will keep moving. Therefore, in such cases, "to *stop* is to get *stuck*." Yet, on a flat and even pavement, a horse will *start* any load he can keep moving *when* started.

In this respect, the Belgian, though not as objectionable as the cobble pavement, is, nevertheless, very defective; as will be seen in Fig. 2 (page 21). Even the Russ, in Broadway, New York, is not free from this objection; while it will be seen, by referring to Figs. 3 and 4 (pages 23 and 24) that the Nicolson is *perfect*; there being no point of its surface more prominent than another.

3d Reason why a flat and even surface is necessary is, to make riding Easy, Agreeable, and Safe.

As before implied, the aim is to have all vehicles move as nearly as possible in a horizontal plane; and were a mathematically perfect *wheel* and *plane* practically possible, there would be no *up and down* vibration to the axle, and such a result would be easily and perfectly attained. Yet, wagon wheels, to all intents and purposes, may be considered *perfect.* But the imperfections in the surface of roads and pavements, producing so much vertical vibration, diminished the pleasure of riding to such an extent as to lead to the invention and introduction of *springs* to wagons; that is, for pleasure wagons, at first. But for other reasons, as will be seen further on, they finally came to be applied to all kinds of vehicles.

At first, then, wagon springs were invented for ease and comfort, and were considered a *luxury*, and therefore, extra toll was charged for spring wagons, while it was admitted they did less harm to the road. Well *may* wagon springs be considered a luxury in these days of cobble and Belgian pavements. Though the former is the most disagreeable of all, to ride over, yet even the latter, compared with the ease of riding on the Nicolson, is *terrible.*

It may be said that *springs* render these pavements tolerably comfortable. This may be measurably so, with light wagons on very elastic springs, or heavier vehicles mounted on compound springs; after the method of the New York omnibuses. But such are only a small proportion of all vehicles; while the light vehicles, after all, seek to get off of the pavements, whenever pleasure and comfort are the objects of driving.

It is true, the surface of the Russ, in Broadway, New York, is tolerably flat, but a different defect in that, (viz.: want of foothold,) together with lack of room, renders pleasure driving, there, not only disagreeable, but quite impossible.

But admitting that springs *do* diminish the rough ness of the cobble and Belgian a trifle, yet the relative difference of ease and comfort, in riding over the various styles of pavements, still holds good: that is, if the use of springs, on the cobble stone, changes perfect torture to common misery; and, on the Belgian blocks, diminishes ordinary misery to bear toleration; then, on the Nicolson, it must render what was easy and comfortable, positively *delightful.*

It is a waste of words to say more on this point: yet, if there is any person *so dull* as not to *see* the truth of these statements, it will only be necessary to put him into a cart, with or without springs, or into *any* springless wagon, and drive him half a mile over each of these three kinds of pavement, at a lively gait, and he will be *convinced*—

It is doubtful if it would not actually shake to *death*, *kill* half the people of New York city to ride in an unyielding lumber wagon five miles, at a rapid rate, over a cobble stone pavement.

This subject of ease and comfort of riding is one in which all are interested.

There is no harmony in the pursuit of pleasure, when a man spends thousands of dollars for medallion carpets, for instance, to increase the comforts of his home; while both he and his family are obliged to forego the pleasure of riding, unless they first submit to the torturing process of driving from *five to seven* miles, *out*, and as many *back*, to find *three or four* miles of pleasant road; when it is possible to lay a pavement directly in front of his house as agreeable as any park road can be.

Besides, the cobble pavement, and its consequent holes and gullies and mounds, cause accidents and harm, which in many cases would not occur on a pavement of a flat and even surface.

When speaking more particularly of the Nicolson, further attention will be called to the *peculiar* ease and pleasure of traversing *level* and *noiseless* pavements.

4th Reason why a flat and even surface is necessary is, to prevent the Breaking and Wear and Tear of vehicles and injury to Freight.

The wear and tear and destruction of city vehicles are enormous. Any wagon that is made for country use, and, that, on country roads, tolerably free of stones, would stand for many years, would be knocked to pieces in a few months on stone pavements. In fact, vehicles for city use must be made doubly strong and heavy; simply because pavements are not flat and level.

The cobble is much like a country road frozen stone hard in mud-time, on which country people think it quite dangerous to drive faster than a walk.

Countrymen, who, at home, have but indifferent roads, abominate pavements and to save wear and tear go a more distant route to avoid them ; yet for the same reason, they go out of their way to seek a rail track, and change tires and axles to fit the rails.

City makers, of the lighter class of vehicles, will warrant their work for a given time on common roads, but *not* on city pavements.

Pavements are severe on vehicles for the reasons :

1st. By the unevenness of the surface, the felloes and tires become sprung between the spokes, by striking the crowns and prominent points of the stones.

2d. The spokes, tires, springs and axles are often

sprung, strained and broken; in consequence of gullies, mounds and depressions.

3d. The side sliding of vehicles over rough pavements is trying to wheels. On the cobble stone, wheels have no *foot-hold*, so to speak; they strike on short inclined planes, and are continually receiving a kind of short, side tripping, which strains every part of the running gear.

4th. There is a hard concussion between the iron tire and the rough, stony track, which, together with a constant wabbling motion, starts and rattles every joint of a wagon.

5th. The unevenness of pavements increases the wear and tear of tires and runners, those indispensable parts of all vehicles, at least 75 per cent.

6th. The elliptic springs, so valuable to all vehicles, when the draft is applied above them, require a level and flat pavement, free from all holes, gullies, and abrupt obstructions, or else they are liable to become broken, sprung and detached.

As before intimated, springs were first applied for ease and pleasure; but it was soon found that they were not only agreeable, but economical; for wheeled vehicles, of every description, will stand much longer when mounted on springs. On city pavements it would be impossible to dispense with them. Even the heaviest and coarsest trucks employ them—unless it be coal and dirt carts.

In fact the roughness of pavements would ruin many kinds of freight, were it not for springs. Yet a smooth and *suitable* pavement would accomplish far better results *without* springs, than the best stone pavement, *with* them.

Freight can be transported and vehicles run, upon

the Nicolson with more ease and safety, and with less wear and tear, without springs, than on the cobble or Belgian with springs.

5th Reason why a flat and even surface is necessary is, to prevent Wear and Tear to the Pavement itself, whatever be its mode of construction or the nature of its material.

A flat surface ensures an even and steady pressure of the vehicle upon the pavement, while an uneven surface produces a hammering action, which diminishes the pressure on some blocks or pieces and intensifies it on others. It is this hammering, battering and ramming down action that settles and displaces the pieces, and produces holes and gullies. It also destroys the immediate surface of each block, by hard and disintegrating concussion.

This is well illustrated by the action of car wheels on their track. Though the rail is straight and smooth, and the wheel perfectly smooth, true and circular, and both made of iron; and though the ends of the rails are on a common level, and rest in iron beds, and firmly wedged and spiked down; yet, the very little jar, produced by the wheels passing from one rail to another is sufficient to make it impossible to prevent the rails from being battered and bruised out of order and place; while the balance of the rail, where there is no jar, remains in perfect order for an unlimited time.

If this be the result on a level iron track, so firmly supported and bound, how much greater must be the injury to a stone pavement, universally uneven, and unsupported by any substructure or superstructure, with the pieces simply laid on the ground and partially rammed down.

In this respect, again, the cobble, Belgian, and Nicolson hold the same relation to each other. That is, the cobble is not only worse than the Belgian, but the worst that can be invented. For, besides being uneven as possible, its pieces have but a single point of contact with the wheel, no flat base to rest upon the ground and no plain sides to support each other in a lateral direction.

By referring to the diagrams, Nos. 1, 2, 3, and 4, again, it will be seen that in this respect the Belgian, over the cobble, is quite a step in advance. Its pieces approximating the form of a cube, though irregular in form, size, angle, and sides, secure a more uniform surface for the wheel, a broader base to rest on and a better lateral support.

But the Nicolson, in this respect, having a flat surface for the wheel, supported with an adequate substructure, and its pieces having secure lateral support, is a complete success; as will be seen when speaking more particularly of this pavement.

6th Reason why a flat and even surface is necessary is, to Facilitate the Cleaning of Streets.

Whatever there may be, connected with pavements, that bears upon the cleanliness of a city should be well considered; for the importance of keeping streets clean cannot be overestimated.

Among the many reasons why streets should be kept free as possible from dirt, are :—

1st. *To prevent dust in dry weather.* The nuisance of dust, at some seasons of the year, is almost insufferable ; besides being injurious to the eyes and health, destructive to clothing, dry goods, and merchandise generally.

2d. *To prevent mud in wet weather.* Mud is also destructive to clothing and often proves a source of great annoyance, by being spattered, by horse and vehicle, in all directions. Besides, mud renders the pavements, in the streets and on the side-walks, so slippery as to make it dangerous to both horses and men. On flat smooth pavements, like the Russ, in Broadway, it is almost impossible to cross the street, among the interminable tangle of vehicles, guided by reckless "Jehus," without hazarding limb and life ; while women and children are often compelled to pass up or down several blocks to find a crossing at all ; and then, likely, be compelled to wade, ancle deep, through slush and dark waters.

Unless the streets are kept clean there is a universal stir of dirt. Mud is tracked from street to side-walk, from side-walk to house ; when dried it is swept to the street again. Dust enters every door, window, and crevice, lodging on every object and then again is driven into the street.

3d. *Streets should be kept clean to avoid pestilence and disease.*

It is well known that the condition of the streets, as regards cleanliness, has much to do with the general health of the community.

4th. *Streets should be kept clean to prevent the mixture of dirt and snow.*

To remove snow from all the streets of a city, by mechanical or artificial means, is impossible. Yet snow is the greatest obstruction to travel, especially when mixed with dirt. The mixture of snow and dirt, packed and rammed down, forms mounds and hummocks as resistant and unyielding, to horse and vehicle, as so many rocks, almost defying the force of pick and axe; and often accumulating, one layer upon another, until they resist the power of sun and atmosphere from the first fall of snow until late in the spring; thus presenting powerful resistance to the movement of vehicles, and weakening the power of horses, by almost entirely depriving them of their foot-hold.

Yet any fall of snow, if perfectly free from dirt, would mostly melt and pass away, in this latitude, by two or three days' action of a warm sun and the salt humid atmosphere of the surrounding ocean.

As the cleanliness of streets, therefore, is of so much importace, and the facility of keeping them clean so much depends upon the nature of the face of the pavements, just refer back to the diagrams Nos. 1, 2, 3 and 4; which represent the surfaces of the cobble, Belgian and Nicolson; and it will be seen that, in this, also, as well as other respects, the cobble is the *poorest* and the Nicolson the *best* pavement; while the Belgian, though superior to the cobble, is far *inferior* to the Nicolson.

Sweeping or cleaning, by any method, the cobble and Belgian pavements, is no more feasible than sweeping, with a common broom, dirt and sand from a house floor that is filled with half-bored auger-holes.

Streets will never be properly cleaned until they are systematically and regularly swept. And this will never be done until it is accomplished by machinery; which is not possible except the pavements be flat. This has been shown by experiment. Sweeping machines that worked well on the Broadway Russ, would not operate, at all, on the cobble.

Compare the operations of an almost impassable gang of men, armed with hoes, shovels and brooms, working days and weeks, even, in trying to clean a single street, by picking dirt and filth, with the sharp corners of their instruments, from the innumerable deep, triangular—non-get-at-able crevices in a cobble stone pavement; and yet, leaving the work so imperfectly performed that you would scarcely know it were done at all. Compare this, we say, with the performance of a powerful sweeping machine, marching once up and down a thoroughfare and leaving it as clean as a swept house floor.—Were flat pavements universal, sweeping machines would be in common use; for inventors will produce what is *possible* and needed.

Besides, streets, as also pleasure roads, like those of Central Park, should be swept instead of sprinkled. **1st,** Because sprinkling will produce more or less fine mud, which will be thrown by the wheels, besides somewhat destroying the foot-hold of the horse—**2d,** Because dirt and street accumulations, being allowed to remain, however much sprinkling may be done, become cut, triturated and dried, when they will necessarily evade the broom and shovel—**3d,** Because machine sweeping, not requiring daylight, enables the cleaning to be done in the night time, when the streets and roads are not required for business and pleasure.

Besides, the dampness of the night prevents dust from flying, which increases the chances of collecting it; and, further, the shops, being closed, are not disturbed by it.

7th Reason why a flat and even surface is necessary is, to Increase the Capacity of Thoroughfares.

By having a level and flat surface the speed of

vehicles is increased, as well as the power of the horse to draw, and the vehicles to bear greater burdens.

Besides, it renders all parts of a street available, by leaving the driver free to seek his way between and among other vehicles, instead of giving his attention to steering, *here and there*, to shun holes, gullies and ridges.

8th Reason why a flat and even surface is necessary is, to render Walking more Agreeable, and so the Streets more Available to Pedestrians.

Every person, on foot, is obliged more or less to cross the streets ; but walking on cobble stones is next to impossible; at any rate, it certainly is very disagreeable, and trying to the feet and ankles. An acknowledgment that they are unfit to walk on is found in the crosswalks, made of flat stones, placed at the corners. It is much more convenient, many times, to cross in the middle of the block, to reach destination sooner, and to better avoid the rush of vehicles. Although the sidewalks are intended for pedestrians, yet the amount of street crossing is immense.—For every block walked, a street must be crossed. Many times, at certain points, the sidewalks could, and would, be relieved by people walking *in* the street between the curb stones, were a perfectly flat and dry pavement, like the Nicolson, laid in all parts of the city. Such a pavement, by affording universal crossing and sure foot-hold, would enable persons to pick the best opening among vehicles and to avail themselves of it, to save the risk of losing life and limb.

9th Reason why a flat and even surface is necessary is, to Diminish Noise.

The noise on a flat pavement, though made of the

same materials, will not be so great as on a rough and uneven surface. Hence the noise of the *cobble* is greater than that of the Belgian, and the *Belgian*, greater than the Broadway Russ; while *either* is incomparably more noisy than the Nicolson.

As the quality of *noiselessness* is of such paramount importance, and the Nicolson, in this respect, standing, as it indisputably does, pre-eminently superior to all others, the pleasure and advantages of this element in pavements will be considered when the peculiar merits of the Nicolson come to be more fully described.

10th Reason why a flat and even surface is necessary is, to make it Conform to the Foot of the Horse.

No desire or skill of man can change the foot of the horse. Its form and nature are unalterably fixed. Therefore, the pavement must conform to the horse and not the horse to the pavement.

The shape of his foot is flat on the bottom, and made to come into contact with the earth on the outer portion of the hoof, while it is absolutely indispensable that the *frog* of the foot be protected; which is accomplished by providing a flat and even pavement.

Here, again, the cobble stone pavement is shown to to be not only a defective and absurd device, but an invention to inflict untold suffering and cruelty upon the noble horse.

The oval surface of the stones, and the deep depressions formed between them, make it impossible for the horse to find a flat resting place for his foot, or any two like places, or uniform surface, upon which to tread with assurance and confidence. Thus his feet are being continually wrenched from side to side, in every direction ; while the upward oval projections of the stones are continually pressing against, and bruising the frog of the foot ; on which *no* pressure should be allowed ; which is not only inhuman to the horse, but destructive of his usefulness to man. (See figure 7, page 41.)

11th Reason why a flat and even surface is necessary is, to prevent the Horse from losing his naturally free and mechanical Gait.

The free and easy stepping of a horse depends much upon the assurance afforded him by a firm, level and uniform roadway. He soon acquires the muscular habit, or habit of motion, adapted to the thoroughfare on which he travels. But, where there is no *uniformity* of surface, he acquires no feeling of confidence, or habit of an easy, mechanical movement.

Notice how insecurely a *man* steps on a cobble pavement, and how he hobbles along, instead of walking with a firm and true step. His gait is entirely changed, simply by coming from country paths or going from flat sidewalks to the rough city pavement. It is just so with the horse. *His* gait is not the same, nor *can* it be, on a rough uneven surface, that it is on a firm level track.

This may not be observable to some persons, yet every jockey *knows* it, and prefers a country road when he desires to exhibit the movement of his steeds to the best advantage.

So far, we have spoken only of *flatness* or *evenness*, which is one of the three requisites of a good pavement that appertain to its surface. The next requisite for consideration, relating to the surface, is *Foot-hold.*

Second requisite of a good Pavement is, to have it so provided with Grooves or Cells as to afford a sure Foot-hold for Horses.

The several reasons, given in the foregoing pages, why the face of a pavement should be flat and even, would also indicate that it should be one *continuous* and *unbroken* surface. And so it ought; were it not that the horse must have a stronger foot-hold, than would be afforded on such a surface by the mere consequent friction of his weight. But iron shod horses, on such a surface, made upon any material that will stand city use, cannot obtain the necessary foot-hold for the various needed purposes of speed and draft, except by means of heel or toe corks, or both. And, as these cannot penetrate the solid substance of the pavement, it must be provided with suitable corresponding depressions, grooves, channels or cells to receive them.

So that, while for all *other* reasons, (some of which appertain to the foot of the horse itself), a flat and continuous surface is needed; it becomes, for *this* reason, indispensably necessary to *break* and *cut* the *general* surface into many *small* surfaces, with these innumerable cells or grooves.

Though these grooves, being small, may seem subordinate to other considerations; yet, even as the flat surface, itself, cannot be employed without them, they become, even to *it*, as well as to the horse, absolutely indispensible. Therefore, the *manner* of *making* them is of no little account; as will be seen :—

1st. *Their direction.* As it is principally the starting, drawing and stopping of vehicles that make the demand for sure foot-hold; and as the wheel in its diminutive dimension, (which is the width of the rim or tire,) would, otherwise, have a tendency to work into them, they should be made transversely to the street. In other words, they must run *across* the line of travel; and extend from curb to curb.

2d. *They must not affect the general surface.* To prevent them from effecting the general flatness of the pavement, they must not be so wide as to allow the periphery of the wheel, in passing over them, to sink, at all, below the general surface. Hence, they should be no wider than is necessary to admit the corks of the shoe.

3d. *Their depth must equal the durability of the pavement.* That is, they must not only remain until the pavement is otherwise worn out, but they must retain their *perfection* and *efficiency*, from the time the pavement is laid until it is taken up. Hence, they must not be mere shallow grooves, only *slightly* lower than the general surface, like those that *once* existed on the Broadway Russ; or mere accidental projections and depressions, like those of the Belgian, which continually diminish by use; but they must be made in the form of cells, as deep as it is intended that the pavement shall ever wear down.

4th. *Their distance apart.* They must not be nearer to each other than is required by the foot and shoe of the horse; otherwise the general surface of the pavement, and its durability would be unnecessarily diminished. Neither must they be so far *asunder*, as to allow the foot to slip before the corks of the shoe enter them.

5th. *Their form must be adapted to the form and action of the corks.* The corks forming right angles with the face or bottom of the shoe, and their four sides being perpendicular, determine that the grooves or cells should have perpendicular sides, forming right angles with the general surface of the pavement; so that, however great be the speed or draft of the horse, the corks will not slip out of the grooves.

6th. *Their distance apart should be uniform.* The free and natural gait of a horse does not depend wholly upon a flat surface and foot-hold, but upon his assurance of *finding* the foot-hold *just on the spot wherever his foot may fall.* Or, as, on city pavements, this is not quite possible, the grooves should not only not be too far apart or too near together, but they should be made a *uniform* distance from each other; so that the slight unavoidable sliding of the shoe-corks into the grooves may be uniform, or not greater than a slight given distance; which enables the horse to acquire almost as much feeling of assurance and freedom of action as he would on a good country road.

A horse submitted to uncertain foot-hold and indefinite sliding soon acquires a mixed and hobbling gait that *destroys* his natural action.

An old Broadway stage horse would find it difficult to speed off freely on a country road, while a country road horse would find it impossible, at first, to take the place of the old stager.

Foot-hold on the Cobble Stone Pavement.

These general principles, relating to the grooving of pavements are grossly violated by the cobble pavement, in that that the depressions, instead of being such grooves or cells as described, are cavities the tops of which wholly occupy and so totally destroy the general surface, both for wheel and foot, and yet entirely fail as a foot-hold. The sides of the stones form rounded inclined planes, on which, if the speed or draft be great, the corks will not catch or fasten, however heavy be the horse.

This is illustrated by the following diagram, which explains itself.

Fig.7.

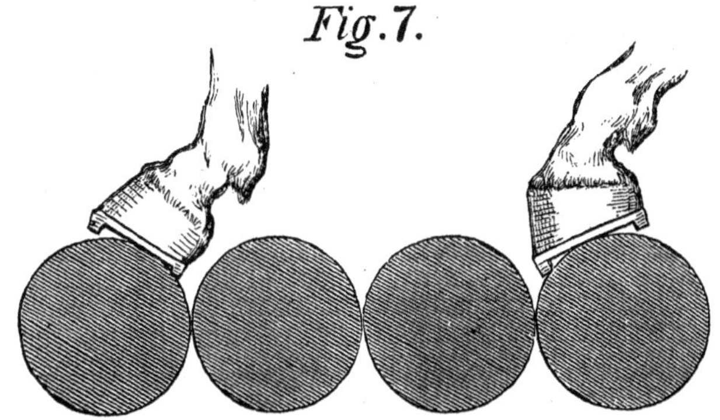

Foot-hold on the Belgian Pavement.

Though the irregular depressions, and the seams between the blocks, of the Belgian, afford a better foot-hold than the cobble, with less unevenness of the surface, yet these seams and depression, not conforming in shape and size to the corks, come far short of providing such a *sure* and *reliable* foot-hold, as speed, draft, safety and ease of travelling require. This will be more readily seen by refering to fig. 8 ; in which the grooves formed by the seams, being triangular (at the top) will allow the cork, when acted upon violently, to slip out. The seams being too narrow to admit the corks are of but little or no use.

The general unevenness or roughness of this pavement, which, at first, aids in affording foot-hold, after

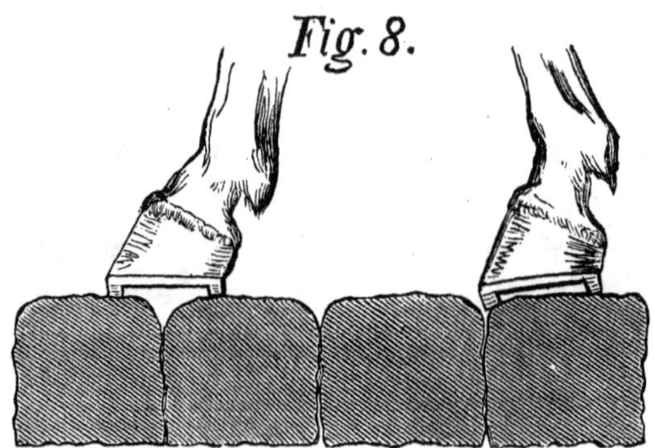

a while becomes hammered and worn off; leaving the face of the blocks very smooth and somewhat rounded; so that, were it not for the nearness of the seams to each other, it would be as smooth and slippery as the Broadway Russ.

A steel file is *best* when *new;* and soon becomes so deteriorated by use that its quality, as a *file*, is destroyed; though its form and substance still remain. So with the Belgian pavement: as regards foot-hold, it is best when first laid; but the face of it soon becomes, like an old file, too smooth to hold, though it be not as flat as the Russ.

It can be observed on any of the pavements that horses, struggling to start heavy loads, stand upon their toes instead of placing their feet flat on the pavement. This is instinctively done to increase the foot-hold; as, otherwise, no hold at all could be secured. Yet it is a fact that, the horse, to exert his full power for draft, must put his feet flat on the ground; which can be done only by securing foot-hold for his heel corks, which is not accomplished by any present mode of paving with stone.

How often the feet of horses fly from beneath them

and come to their places again like cannon balls, to the destruction of their feet and limbs, when a system of grooving like the Nicolson, to catch the heel corks would entirely prevent it.

Should this be doubted, clear ocular demonstration can be had by observing the pavement, and the effort of horses, with heavy loads, on the up grade of Fulton street, between Gold and Cliff streets, New York.

Attention is called to this place because of its convenience, the grade and travel of the street, and the fact that the pavement is otherwise in good condition.

Foot-hold on the Broadway Russ Pavement.

This pavement, though more flat and even than either Cobble or Belgian, furnishes hardly *any* foot-hold. Hundreds of horses slip and fall upon it daily; and, some days, even thousands. In some states of moisture and dirt, (not in freezing weather either,) no strange horse could possibly make his way from Union Square to Bowling Green, at a speed above a walk, without falling several times, at risk of life and limb. The seams are too narrow to admit the corks; and even if they were not, they are too far apart to afford suitable foot-hold.

Besides reasonable flatness and durability, this pavement is a total failure.

Foot-hold on the Nicolson Pavement.

All the foregoing necessary qualities and conditions, in regard to foot-hold and grooving of pavements, are supplied by the Nicolson; as will be shown in the description of its construction hereafter.

The next and last requisite of a good pavement, appertaining to the surface, relates to its general form, which determines the face of the street.

Third requisite of a good Pavement is to have it slightly Arched, and provided with suitable Gutters.

All pavements should be arched, and the oval extended from gutter to gutter.

1st, *To assist vehicles to pass each other without colliding.* A *convex* surface of the street has more tendency than a level plane to throw passing vehicles apart or away from each other; while, on a *concave* surface, there would be a reverse effect.

2d, *They should be arched to facilitate Drainage and Cleanliness of the street.*

For suggestions about gutters see description of the Nicolson.

If the face of the street be rounded and provided with a suitable pavement and gutters, much dirt is necessarily washed off and carried away by heavy rains.

3d, *They should be rounded to increase their Stability.*

With paving blocks of geometrical form, that is, those bounded by right angles, straight lines and opposite parallel planes, the arch can be made so accurately, and the side pressure, thereby, made so great, that no blocks or pieces can be displaced in a lateral direction.

We have, so far, discussed the three requisites of a good Pavement that relate to its surface: namely;

flatness, *foot-hold* and *form of general surface*; which are the *results*, rather than the *construction* of a good pavement.

We now come to the Construction, that is to the *shape* and *dimensions* of the several *parts* and their relation to each other.

Fourth requisite of a good Pavement is, Stability of Structure: That is, its several parts must be so combined together that, when put down, they will Retain their Original Position.

It matters not how flat and even may be the surface when it is *first* laid, unless the general construction of the pavement be such as to ensure its reasonable retention and preservation; for it makes no difference whether the surface soon *becomes* rough, by the blocks sinking and rising, or by being made so originally. It is the same with regard to the foot-hold and general form of the street.

The general construction and method of laying the *body* of the pavement must, therefore, ensure an unyielding and enduring stability.

1st Requisite of Stability of Structure is, a continuous and adequate Substructure.

1st, *Necessity of a Substructure is, an unyielding vertical support of the Blocks, to place and secure them on a common level with one another.*

As no material of a concrete nature can be produced, to form a continuous and unbroken surface, which, for city use, will at the same time resist the action of iron shoes and tires and admit of cells or grooves for foothold, it follows that, whatever be the material employed for the upper stratum, it must be made in blocks or pieces; and as the surface of the ground is not firm enough to give these separate pieces an independent and sufficiently solid bearing, they must, of course, be supported by a suitable and continuous *substructure;* which shall, itself, be supported by the general surface of the street, and thus made adequate to hold the pieces, composing the upper stratum, on a common level. Such a substructure is, to a pavement, composed of pieces, what a foundation is to a building.

The unequaled stability of the ancient Roman pavements is attributable, not to the face-dressing, or the accurate joining, of their polygonal *surface* stones, but to the unyielding substructure on which they rested.

These Roman pavements consisted of three layers. The bottom layer was nine inches thick, composed of stone and cement and constituted the substructure. The middle, six inches thick, consisting of gravel and cement, served to form the union or fitting of the imperfect *lower* surface of the top stones to the rough ragged *upper* surface of the bottom layer. But this middle layer is not needed, when the bottom of the blocks is regular and flat, and the upper surface of the substructure, level and smooth: as is the case with the Nicolson pavement.

The manner of laying the Cobble and Belgian is in direct violation of this method. The ground is simply stirred up and shaped, and the stones laid side by side, without special care as to their exact level with one another, and dirt and sand scattered over and amongst them. These are then brought, approximately, to a common level, by simply driving them with a rammer. Now if a few blows of a mere *hand* rammer will drive them down to a somewhat level position, then, of course, a little longer appliance of the same feeble means, unequally applied, would drive them *out* of position. Hence, how can it be expected that they will resist for *years*, the constant and powerful beating and ramming to which they are subjected. It is simply absurd, to suppose that these bits and blocks of stone, driven into position, as it were, with a blow of the fist or the tap of a hammer, will not be unequally driven still *further* into the dirt, and so *out* of position, when they are beaten, at every turn, with ponderous wheels wreaking and whirling under the pressure of enormous burdens. There is *no* stability to paving blocks not supported with something more than loose earth.

A cobble stone has no base—*nothing* to rest on. Benig oval, it stands on a point. And such a pavement besides being good for nothing, at best (more than to prevent the cutting of the ground,) could be put out of order by beating it, here and there, with a common nail hammer.

The Belgian blocks, being flat on the bottom, rest rather firmer on the ground. Yet, as regards their verticle position, they are not reliable.

2d, *Necessity of a Sub-structure is, to retain the general form of the surface.*

Without a substructure, to serve as a medium of contact between the layer of blocks and the ground, not only will some of the single pieces settle and rise, but *sections* of considerable extent, here and there, will give way and settle; thus forming holes and gullies. These may be seen all about the streets; not only on the cobble and Belgian, but on the Broadway Russ, as well. These depressions occur where the ground, being less firm, allows the blocks to be beaten down. But, were the blocks resting upon a firm and continuous substructure, they could not settle, here and there, wherever the ground might happen to be less firmly packed.

3d, *Necessity of a Substructure is, to assist in rendering the Pavement impervious to Water.*

A pavement, however constructed and of whatever materials composed, should form an impassable barrier to rain and moisture. Otherwise the infiltrations of water from above and the exhalations of moisture from below, together with freezing and thawing, at certain seasons, will more or less heave and disturb the pavement; especially if the blocks rest directly on the ground. Hence the need of a substructure to aid in rendering the pavement impervious to water and impenetrable by frost.

4th, *Necessity of a Substructure is, to prevent the Dirt from working up among and from under the Blocks.*

When the blocks of a pavement rest directly on the ground, the constant action of horse and vehicle has the effect, not only of hammering them out of a common level, but it disturbs the position of the dirt, by removing it from under one stone and driving it under another; thus undermining some and upthrowing others. From the same cause, the dirt works up into the interstices and thus destroys the union of the blocks.

This displacement of dirt, and its consequent injury to the pavement, would be wholly prevented by intervening, between the block and ground, a dirt-tight substructure.

Having considered the necessities of a substructure, we come now to speak of the lateral support of the blocks.

2d Requisite of Stability is an unyielding Lateral Support of the Blocks, to secure them in an upright position.

The lateral support of each piece or block must come from its surrounding fellows. It can come from no other source. Each block must bear upon those next to it, until the curb stones are reached.

In the longitudinal direction one piece rests against another the whole length of the street.

That they do and must mutually support each other in lateral directions is evident; for, if only two or three stones be taken up, here and there, all the blocks around about, will be displaced, and the pavement wholly torn to pieces.

Hence, as regards their lateral stability, it may be said of paving blocks; "*united, they stand; divided, they fall.*"

3d Requisite of Stability is suitableness and perfection of the Shape of the Blocks

Though a pavement is composed of many blocks or pieces, yet, they are but duplicates of each other. And having, in all respects, a common relation to each other, not only must they have uniformity of shape and size, but their form must be such as to afford a plain, uniform and broad-faced contact with each other.

The form, *best* adapted to this purpose, is the quadrate or quadrilateral: the next best is the hexagonal.

Whether the one or the other of these be employed, it is not sufficient that the blocks merely *approximate* such forms, but, (if the quadrate be adopted,) they need not, necessarily, be cubes but they must be parallelograms; that is, bounded by right angles and straight-lined opposite parallel sides; executed with, at least, all possible mechanical accuracy.

There are various reasons why the individual blocks need to be thus uniform and accurately and peculiarly formed.

1st *Necessity of accurate Parallelogramic form of the individual Blocks is, to secure the needed flat and even Surface.*

It being necessary, as before shown, to support the blocks on an even and continuous substructure, in order to retain them on a common level; it, therefore, becomes necessary to have the blocks all equal in their length or vertical dimension, and their face and base parallel and at right angles to their upright sides, that their upper surfaces may lie horizontally and fall into one common plane.

Except by this method there is no sufficiently inexpensive means of producing a perfectly flat and fixed surface.

2d *Necessity of accurate Parallelogramic form of the individual Blocks is, to provide for their secure Lateral Support.*

Since each block, as intimated, must rely upon its surrounding fellows for lateral support, it becomes necessary that the base of each block be made at right angles to its four sides; that it may have a natural tendency to stand in a perpendicular attitude, however great be the downward pressure to which it may be subjected. If the base be *not* flat and at right angles to the sides, then the block, when pressed and hammered by heavy weights, will cant over, in defiance of any possible lateral support, and thus slant the face of the block out of the common plane.

Again, if the sides be not flat and at right angles to each other, it will be impossible for the blocks to afford each other responsive contact and resistance; which will allow them, more or less, to cant over and twist around in their places, and, so, roughen the general surface.

3d *Necessity of accurate Parallelogramic form of the individual Blocks is, to provide Foot-hold.*

Though such form of the blocks is not indispensable to some sort of foot-hold, yet there is no other way of providing foot-hold, that will be so cheap and durable, as by the use of such a formed block; as will be more clearly explained when giving the description of the Nicolson pavement.

4th *Necessity of accurate Parallelogramic form of the individual Blocks is, to prevent the displacement of Dirt and the infiltration of Water.*

If a pavement ought to be, as intimated, impervious to water and resist the action of frost on the earth beneath it; then the blocks should be made with right-lined faces and right angles, in order that the seams between and below them shall be so close that no dirt, or even water, can work into them. The effects of dirt, water, and frost in the interstices of any stone or mason work is always detrimental to the stability of the structure;—especially is this true with pavements, which lie only on the face of the ground and are being continually hammered and beaten.

The Cobble and Belgian pavements are in direct violation of all the foregoing conditions of stability of structure, as is quite evident without comment.

The individual cobble stone has no sides at all, either for base, to rest on ; or face, to be traveled over ; or sides, to receive or give lateral support.

The Belgian blocks, in this respect, though far superior to the cobble stones, are still very defective : for the reasons, that dirt will work between and frost will reach below them. The earth will be displaced beneath them, and work into the interstices. Water will pass between them, causing the ground to settle and let the pavement down in spots. Not being wholly responsive in their lateral bearing and support, they more or less cant over and twist around and leave the surface uneven.

Why have not the ancient Roman Pavements been adopted for Present Uses ?

Having discussed, somewhat at length, *flatness*, *foothold*, *general form of the surface and stability of structure ;* and shown that the Cobble, Belgian, and Russ are each, in some or all of these respects, a failure ; that neither of them will answer the needed purposes of modern cities, the question naturally arises, why not adopt the Roman method, and reproduce their substantial and ever-enduring structures ?

It is true, the Romans had the best pavements ever known, before or since their time—pavements which, for ages, have withstood those elements and forces of destruction that have crumbled to dust all other structures of usefulness and grandeur. Sections of their Appian way even still exist, over which Roman legions tramped more than two thousand years ago.

Though such was the durability of these pavements, yet they would not meet the demand of the present age.

In the first place, were they, otherwise, in all respects suitable, their enormous *cost*, in these days of diffusion rather than concentration of wealth, labor and influence, would render their general adoption impossible.

Again, the peculiar merits of these pavements, consisting only of *flatness* and *stability*, would not answer for modern use ; for, though they possessed these qualities in a remarkable degree, and served admirably the purposes of that age, it must be borne in mind that, horses were not iron shod or wheels iron bound ; shoes and tires being introduced centuries later.

The use of shoes and tires, in some respects, necessitating a change in pavements, this subject will be alluded to when speaking of fitness of materials for pavements.

Besides, the underground net-work of gas and water pipes, in cities of the present day, requires a pavement which will admit of being taken up and replaced with greater facility, than would be afforded by these deep and solid structures.

Why has not the Art of Paving kept pace with Other Improvements?

Streets and roads are a necessity in all civilized countries. All persons come into contact with them, and are ever ready to condemn their bad, and to appreciate their good, condition. Yet, for some reason, their importance has been overlooked, both in city and country. There are, in the metropolis, between curb and cross-walk, sloughs and gutters, (in New York, for instance, around about Fulton and Washington Markets,) filled, ancle-deep, with mud, garbage and foul waters, ever offensive to the senses; and over which hundreds of thousands of people must daily stride; and at the very places too, where, the thoroughfares being thronged with people and vehicles, all attention is required to make a safe passage across the street. Yet these gaps are so easily repaired, that an ordinary farmer would not permit such filthy holes to exist between his house and barn-yard for a week; but our city authorities, aided by dignified, high salaried, civil engineers, allow the befouling pits to be stumbled over and into, day after day, for tens and tens of years. Not only are city pavements miserably defective, as shown on pages back, but the road ways and walks of villages, towns and the country at large, are equally neglected and defective. There is hardly a town or village, in which there are not, at some or all seasons, sandy, muddy, hilly or stony roads and walks, over and through which all the men, women and children, from youth to old age, generation after generation, do not tediously plod their way; when a little united effort would easily convert such places and ways into hard and dry roads and walks, to remain an every-day and perpetual convenience to all.

The reason why so little attention and expense is devoted to the subject of roads, walks and streets is, that, "what is everybody's business is nobody's business." Streets and roads are strictly a public work, of imminent necessity; and, by the Romans, being so considered, they gave to them that strict care and attention which they were wont to bestow on *all* their works of public interest.

In this age and country the matter is reversed;

roads having been improved just in proportion as they have fallen into the hands of private management. In times of stage-coaching, in this country, turnpike roads, which were managed by private enterprise, were tolerable ; but, since the introduction of steamboats and railroads, being abandoned and given over to the public, they have become but indifferent roads again.—Witness the development of travelling facilities under the management of private enterprise, in the construction of railroads, river and ocean steamers, &c.—a development which, in boldness of undertaking, cost and success, and general benefit to the community at large, is completely astounding, and astonishes the world ; and challenges comparison with any Roman enterprise, private or public, and, relatively, converts her famous Appian way into a mere cross road or byway.

What the Romans did, for convenience of travel, by public, we do by private, enterprise.

Paving of cities, however, is still managed by public authorities, and behold the miserable streets. House and store-building is done by private management, and witness the splendid erections.

Of the four sides of a street, heaven displays above its soft blue sky and golden sun ; on either side private enterprise displays its unbroken lines of magnificent structures of usefulness, so decorated with taste and ornament as to tempt the camera and command the admiration of the artist—and on the *bottom*, the city government tumbles down a rough layer of cobble stones, filled with holes, gullies, pits and sloughs, dirt and mud, for horse and man to hobble over as best they can.

Why such large, flat and costly flag stones in front

of dwellings and stores, and such mean apologies for pavements? Because the curb stone draws the line between private and public enterprise.

If the property owners, on some principal street of New York, were to have the privilege, for all time to come, of selecting and paying for such pavement as they might prefer, they would invite attention, encourage invention, try experiments, make progress, and soon have down a better pavement than the city authorities, backed by professional engineers, ever dreamed to be possible.

Popular Prejudice and Opposition to new Ideas, Modes and Things.

To select the best means to produce desired results, in the mechanic arts, constitutes invention, and invention implies something new; that is, new ways and means of doing the same thing. And the only idea we *have* of perfection, in this department of human affairs, is, that nothing *yet* known or invented is *more* perfect. Hence, all things mechanical may be still further improved. And the natural tendency of many minds is, to forever seek for new and better ways and means of accomplishing desired results. And were not this live spirit of discovery and invention staid and counteracted by the dead weights and forces of popular ignorance and prejudice, selfish opposition and almost hate, with here and there a few wiseacres, sitting in high places, refusing to open their eyes for fear they *will see* something—declining to look at the new moon out of respect for the old, there is no calculating what *might* have been the progress of civilization at the present day.

New inventions and ideas of sufficient magnitude to attract general attention always meet with this great deaf and blind spirit. Mankind, not only the unenlightened populace, but the magnates in authority, have carried this opposition and prejudice so far as to persecute their greatest inventors, discoverers and benefactors; rejecting the most important improvements for many years.

They would not be told, nor much less believe, the world was round and turned round, but, still preferring to believe it was motionless and flat, persecuted Galileo for discovering the great facts; they would not encourage Columbus; pursued Harvey and harassed him from youth to old age for telling them of the circulation of the blood; persecuted political and moral reformers, generally; would not believe in railroads, steamboats or steam; pronouncing them works of the devil. The Paris Academy of Sciences early in this century, being consulted by Napoleon on the subject of steamboats, pronounced them, "*a mad notion, a gross delusion, an absurdity.*" New York newspapers, in the *nineteenth* century, even after reluctantly admitting they might do for freight, seriously argued that they would *never* do for passengers, as "*the wheels would spatter water so.*" And even after steamboating had come to be a magnificent success for years and years, it was still contended, by even noted engineers and philosophers, that an *ocean* steamer, for a hundred and one good reasons, was all moonshine, and utterly impossible. They disbelieved in, and opposed the use of iron ships, the telegraph and the sewing machine; opposed the introduction of new articles of food, as potatoes, tomatoes, pieplant, &c.; even during its late famine starving Ireland resolutely refused to eat corn meal.

In short, *all new* modes, things and ideas are, at first, rejected and fought against; but gradually the darkness gives way, the light shines, and improvements of true merit find approbation and come to be universally indispensable. Civilization refuses, at *last*, to part with what she had *once* rejected and condemned. Her tenacity for things *old*, is only equaled by her opposition to things *new;* she first condemns; then, admires; stigmatizes, then glorifies.

Yet the world *is* round and it *turns* round; the Western hemisphere *does* exist; and the blood *does* circulate; reformers *have* gained their points; steam coaches *do* whiz and thunder over hill and plain; and steamboats *do* plow the river and the ocean; the "Great Eastern" *does* float; the telegraph *does* flash our thoughts with lightning, even through ocean depths, five thousand miles per second; and the sewing machine *does* click its ten thousand stitches a minute; and even Ireland *does* eat corn meal; and the world is glad; as also the people will hereafter rejoice over the Nicolson pavement.

Hence it is not strange that prejudice and opposition should be shown to the introduction of a new material for pavements; especially when the new is as unlike the old, as wood is different from stone.

Yet, as regards the Nicolson pavement, its advantages are so striking, every person who has rode over or seen it, at once, admits and exclaims, that is just the thing needed and desired if it will only *stand.*

But since the durability of anything is not tested by looking at it, and wood being fibrous, and so unlike stone, and, as pavement, being submitted to iron shoes and tires, it is not strange that the question of durability should be raised.

But we think it can be demonstrated to any unprejudiced mind, that wood, taken in all respects, is the best adapted material for pavements yet known. And as soon as its merits, for this use, become more widely known and appreciated, it would not be surprising to find it rapidly superseding and subverting all other paving materials ; not only in the West and other places, as it is, but everywhere.

The failure of other Wooden Pavements, than the Nicolson, not owing to defects of Wood as Material, but to the Inappropriate Method of Construction.

Wooden pavements, like everything else, can be poorly made. And so, too, can the Nicolson. For success or failure often depends, not alone upon the particular *general* method, or kind of material employed for a given purpose, but upon the *exact* manner of arrangement, appliance and perfection or imperfection of execution. Bad workmanship and poor quality, even of the same material, will spoil anything. For instance no one questions the general adaptation of wood, as a material, for household furniture. Yet, for given articles, the wood must be the right kind, good quality and well seasoned, the construction suitable and the workmanship good ; or the articles shrink to pieces come apart and break down.

Success or failure often depends on the *mode of appliance ;* as, for instance, a chair leg, made crossways of the grain of the wood, would be good for nothing. So, a farmer's beetle, used on its sides would soon fail, but apply it to the wedge endwise of the grain and it endures for years. A knife must be applied edge to the wood.

In machinery, sometimes failure or success will depend upon *a drop* of oil; hence the saying, "oil is the half of a machinist." The linch-pin of a wagon is a small matter, yet it, or its equivalent, is as important as the wheel itself. A *screw*, missing, may cause a railroad disaster, or the loss of an ocean steamer.

In chemistry, exact kind and quantity are indispensable to success. It will not do, for whatever reason, to say, this instead of *that*, or a little *more*, or *less* of *this* or the *other* ingredient will *do;* for, no matter how slight be the deviations, the results are as entirely different and the failures as absolute, as if the deviations had been ever so great.

In medicine, different, and even opposite results are produced by giving the *same* medicine in different quantities; or in giving the same quantities under different symptoms of disease.

So, in short, to ensure success, the laws governing mechanical matters must be as implicitly complied with as those of mathematics, chemistry or any other science.

So with wooden pavements. Prior to the Nicolson, though the blocks were set on end, yet the structure did not afford a *flat surface* or *foot-hold*, or protection from decay; either one of which defects would be sufficient cause of failure. Not being protected from decay, the wood became soft and tender, and so would not stand the action of shoe and tire; and not being grooved, became slippery.

Why the Nicolson has sometimes failed.

The Nicolson, itself, though adequate to meet all the requirements of a good pavement, has, in some in stances, been laid in such violation of the prescribed rules and principles of its construction, that it could hardly be considered the Nicolson at all.

In some places, where it has been laid on new-made ground, it has settled; as, of course, it would ; for it is not expected, of any pavement, that it will support itself in the air without resting on the earth.

Therefore, when we come to explain the advantages of this pavement, it is to be borne in mind that, we refer to such a specimen of pavement as answers to the prescribed rules and modes of construction as are given on the immediately following pages.

DESCRIPTION OF THE NICOLSON PAVEMENT.

1st. Preparation of the Ground.

The ground is leveled, or rather rounded off, in such oval form as, by situation, width of street, &c., may be required ; after which, in order to procure nice uniformity, it is thoroughly raked and made as smooth and even as a house floor.

2d. The Substructure.

This consists of *two* inch white or yellow pine planks, laid side by side, lengthwise of the street, so as to entirely cover the ground. The planks are then completely covered, by means of suitable brooms, with a coat of liquid asphaltum and coal tar. The planks are now turned, one by one, the other side up, and treated in the same way with the asphaltum and tar, and allowed to remain. This constitutes the substructure, which is now ready to receive the upper stratum.

3d. The Upper Stratum.

This consists of white or yellow pine parallelogramic

blocks, *eight* inches long, (with the grain of the wood,) *three* inches thick and from *six* to *ten* inches wide, with straight-lined faces and right angles; together with strips of board, of the same material, *four* inches wide, *three-quarters* of an inch thick and *five* feet, or so, long; laid in the following manner. Beginning, for instance, at one end of the street, a series or row of these blocks is set up on end, across the street, from curb to curb, with their broad faces fronting up and down the street. But before thus placing them in position, each block is completely coated with asphaltum and tar cement, by being submerged in it. The first line of blocks being thus set, a line of strips, a section of which is shown by F (figure 4), also coated with the cement, is laid and nailed up against the blocks D D with one edge resting on the plank substructure E E. Now another row or series of the blocks, treated in the same way, is set up against the strip; and so on, alternately; until several rods, more or less, are laid.

There is now left between each two consecutive rows of blocks, a continuous groove or cell, *three quarters* of an inch wide and *four* inches deep, extending from curb to curb, which constitutes the foot-hold.

4th. The Filling.

This consists of clean gravel, screened so as to vary from the size of a *pea* to that of *five-eighths* of an inch in diameter; together with asphaltum and tar; treated and applied as follows: The gravel is heated very hot, in suitable sheet-iron pans, and then filled into the cells level with the surface. Asphaltum, also heated hot, is now poured, (from a vessel provided with a spout,) into the cells until they are filled with it, and while the gravel is yet hot. And, (before cooled,) this filling (of gravel and asphaltum) is swedged or rammed; which is done by means of a bar of iron of suitable width, *two* feet, or so, long and just thick enough to enter the cell, which is placed edgewise on the filling and hammered upon with a heavy rammer. The whole surface is now treated to another finishing coat or covering of the asphaltum and coal tar.

5th. The Dressing.

This consists of a layer of sand and finer gravel, spread over the entire surface, about an *inch and a half* thick, and allowed to remain; which becomes partially pulverized and ground into the fibres of the wood, washed and worn off by the action of rain, horse and vehicle.

The pavement is now complete.

It will be noticed that, to firmly secure and support the several parts and pieces, they are so arranged and laid as to break joints; and that, while the intervening strips separate the blocks at the top, to furnish foothold, they also serve to unite them at the bottom in such a manner as to make them self-supporting, and render the upper stratum, (consisting of the blocks and strips,) one continuous and substantial body of wood.

The following cuts, drawn on a scale of an inch to the foot, represent views of the Nicolson pavement, as seen from different directions.

Illustrations of the Nicolson Pavement.

Figure 3 represents a longitudinal section, as viewed from the sidewalk. E E shows the substructure or planks; D D D, the blocks; F F, the strips; G G, the filling. The dressing is not shown.

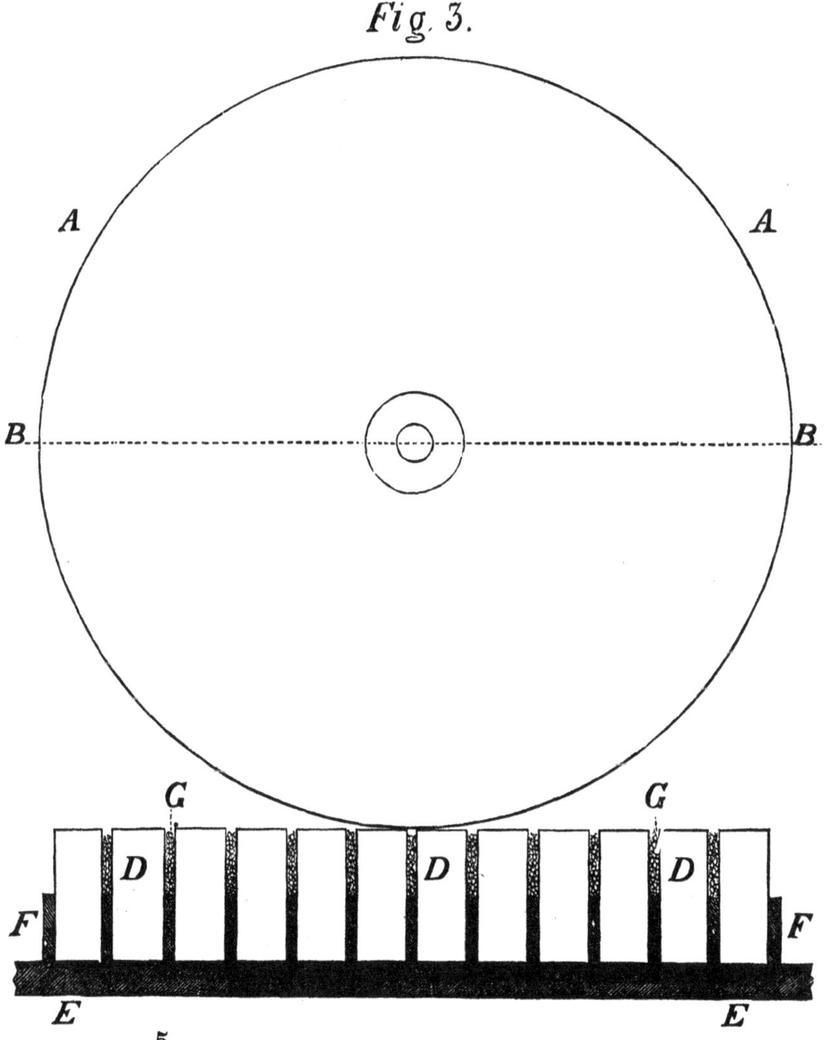

Fig. 3.

Figure 4 represents a section as viewed from up or down the street; E E showing the ends of the planks forming the substructure; D D, the blocks, seen on their broad sides; F, the strip, that separates the different series of blocks.

Fig. 4.

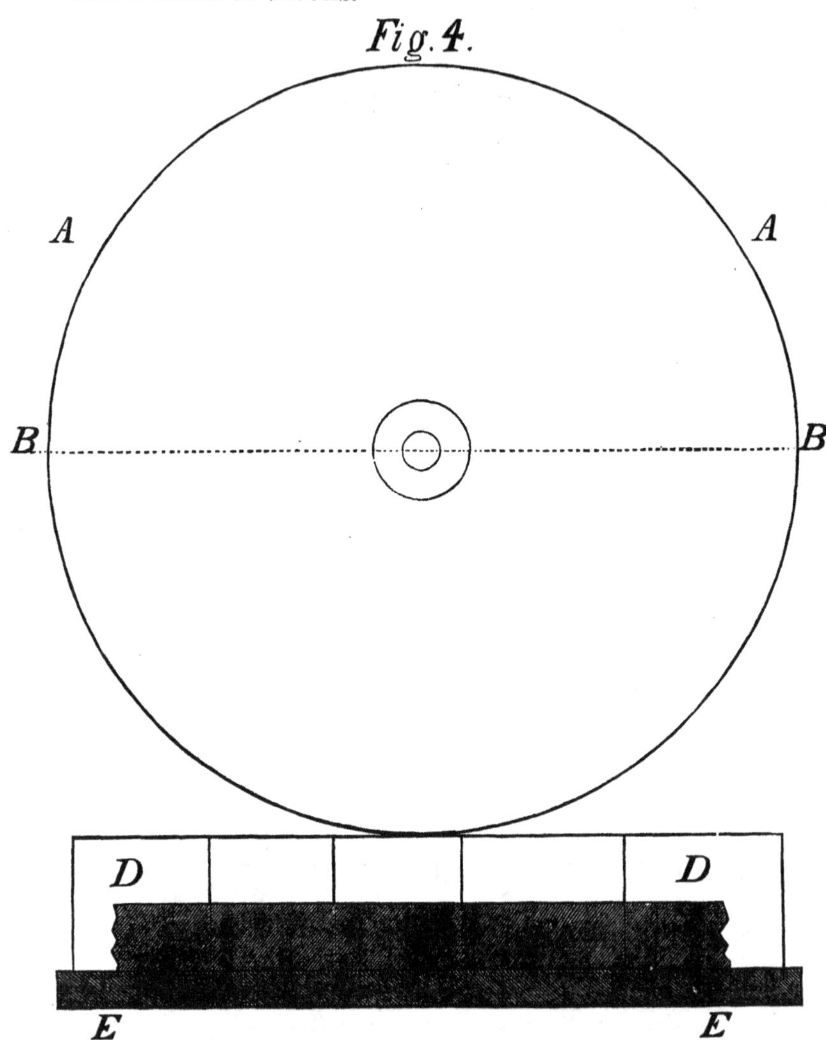

Fig. 5.

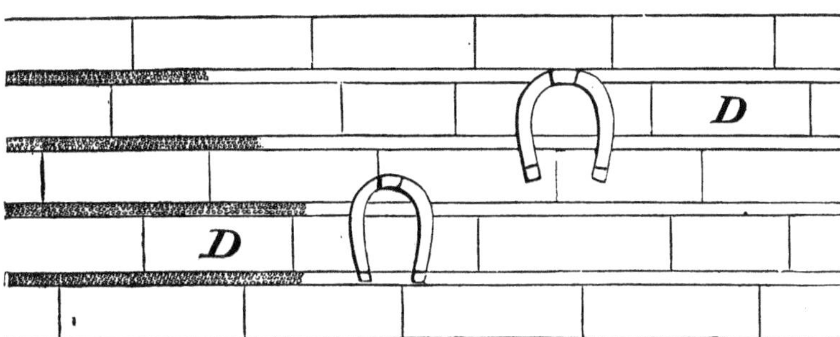

Figure 5 represents a face view, as it would appear when seen from above, after the dressing is worn and washed off—the filling being left out on the right.

Figure 6, (see following page) drawn on a much smaller scale, represents a transverse section extending from curb to curb, as seen from up or down the street; the "strip" (F fig. 4) not being shown.

The curvature is over drawn, as the street would not require to be so rounding. C C are the curb-stones; E E, the substructure. The two blocks, D D, are termed the gutter blocks; the face of which slants up (instead of down) towards the curb, which brings the bottom or deepest line of the gutter a few inches off from the curb; the advantages of which will be explained further on.

The objects of the several parts and their general arrangement will be fully explained in describing the advantages of this pavement.

Fig. 6

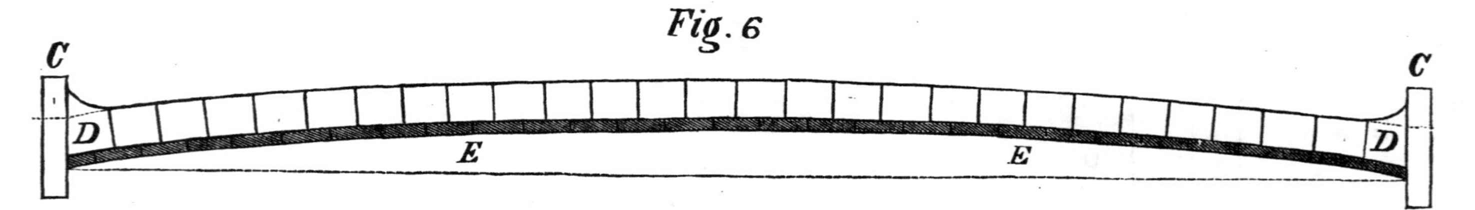

ADVANTAGES OF THE NICOLSON PAVEMENT.

Having previously described, somewhat at length, the requisites of a good pavement, irrespective of the material that might be employed, and shown that neither the Cobble, Belgian nor Russ supplies these requisites ; we now propose to show that the Nicolson not only meets all *these* requirements ; but that, by the superiority of wood over other materials, it accomplishes *more ;* in that, that it constitutes a pavement relatively *noiseless*, and infinitely better adapted to the nature of the Horse and the action of iron shoes and tires than any mineral or metal pavement can be, however constructed.

In describing the advantages of the Nicolson, as they relate to the requirements already discussed, it will not be necessary to much more than call attention to the construction of the pavement in connection with a bare enumeration of these requisites, to show that it supplies all the conditions necessary to insure perfect success.

But in discussing the superiority of wood over stone, as relates to noise, durability, adaption to the nature of the Horse and iron shoes and tires, we shall undertake to show, that heretofore, a great mistake has been committed in the selection of *materials* for pavements.

First. Advantage of the Nicolson Pavement as relates to a flat and even Surface.

1st. The Surface is Perfect.

By means of the substructure, which is as smooth and uniform as a house floor ; and the exact uniformity of the length or height of the blocks, together with the fact that the face of each is a perfect plane, renders the general surface of the pavement perfectly flat ; so much so, that the sweep of a straight edge or line would touch upon or stroke the surface of every block. Yet this perfection of the surface, (the first of the three cardinal points of a good pavement), is obtained without depriving the horse of his sure foothold ; while in other pavements, though the *surface* is often sacrificed to secure foothold, or foothold to obtain surface, yet neither is obtained with any perfection ; as is illustrated by the Russ and Cobble pavements.

2d. The perfection of its Surface conforms to the perfection and nature of the Wagon Wheel.

Wheels, on the theory of their rims being everywhere equally distant from their hubs, rolling over this pavement, carry their axles in a horizontal plane ; while on the Cobble, Belgian and Russ, they throw the axles out of a horizontal plane, in every four to eight inches, as is explained on pages 19 and 21, and illustrated by figures 1, 2 and 3.

3d. Its Surface presents no obstacle to vehicles, and so renders the Draft of Horses more Effective.

There are no points higher or lower than the general surface, behind or into which wheels can be placed ; hence a horse can *start* all he can *draw ;* while it often requires the force of several horses to start a load over the immovable chocks everywhere presented to wheels on any of our city pavements, not even excepting the Broadway Russ.

4th. Its perfect Surface renders riding Easy, Agreeable, and Safe.

That riding over such a surface, in whatever style of vehicle, should be at once more agreeable, easy and safe, is so self-evident that it seems but folly, in argument, to more than say so. Still, when speaking of wood, as material for pavements, we shall allude to the peculiar agreeableness of traversing the Nicolson.

5th. Its flat and even surface prevents Wear and Tear of Vehicles and injury of Freight.

1st. Ninety per cent. of the jar, which so soon starts and rattles the joints and bolts of a wagon, is diminished.

2d. It does away with the side tripping or sliding of wheels, which strains them more than do the burdens carried.

3d. It prevents holes and gullies, which break springs, axles and wheels.

4th. It furnishes a continuous bearing for the continuous rim of the wheel, which prevents the felloes and tires from being sprung between the spokes.

5th. Where springs are applied below the draft, it prevents the peculiar jerky motion of the axle, and wheels which is unavoidable, and injurious on a rough surface.

We shall omit to speak of wear and tear of shoes and tires until we shall have discussed the durability of the Nicolson.

6th. Its perfect Surface facilitates Cleanliness of the Streets.

It renders cleanliness of the streets possible by facilitating the use of sweeping machines; only by means of which will, or can, streets ever be kept properly cleaned. And the surface being oval the dirt is partly washed off by rains.

7th. Its surface increases the Capacity of Thoroughfares.

Its surface being equivalent to a universal rail-track enables vehicles to be run in every direction, thus rendering all parts of the street available. It enables heavier loads to be drawn—admits of greater speed—requires less horses. In narrow and crowded streets this is an important consideration.

Why horse railroads, only that, being smoother, they enable the horses to do more work, as well as promote ease and convenience.

8th. It renders all parts of the Streets available and convenient to Pedestrians.

This and the foregoing enumerated advantages will be more fully appreciated by referring to previous pages, from 17 to 37.

Second. The Advantage of the Nicolson Pavement as relates to Foot-hold.

1*st.* The cells extending across the line of travel, there is no tendency of the wheel, in its diminutive proportion, (which is the width of its rim), to work into them.

2*d.* The united faces of the blocks being continuous

from curb to curb, the wheels, so to express it, always find a foot-hold in a side direction, and that side slipping and tripping, so straining to wheels on a rough pavement, is wholly avoided.

3d. The cells are so narrow that the wheels, rolling over them, are not thrown, in the least, up and down; and the axle therefore is not thrown out of the horizontal plane.

4th. The cells in size, shape, and distance apart are perfectly adapted to the size and shape of the corks of the shoe, and the size of the foot.

The sides of the cells being perpendicular and at right angles to the face of the pavement, and the corks being perpendicular and at right angles to the face of the shoe, there is no tendency of the corks to slip out of the cells until the foot is taken up; because the shoes and the corners of the blocks lock together by corresponding right angles, instead of obtuse angles, as in all other systems of grooving. (See figures 7 and 8, pages 41 and 42.)

This is so clearly illustrated by the following diagrams as to need no further explanation.

Fig. 9.

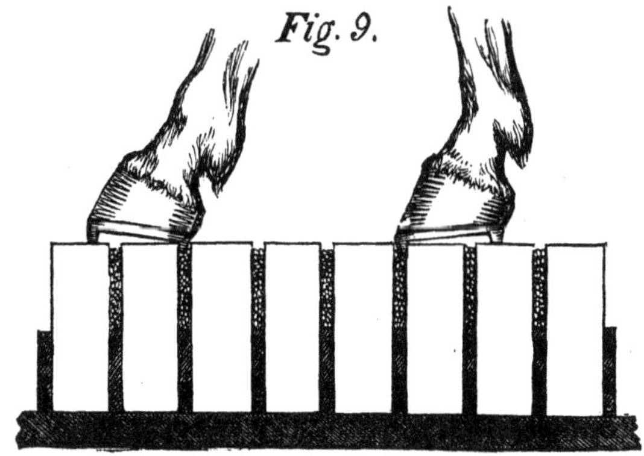

While the width of the cells is sufficient to admit the corks, they afford no unnecessary space to allow slipping after the corks enter them.

The thickness of the blocks, and so the distance between the cells, is uniform, and affords the horse reliance in finding his foot-hold ; while the adaptation, of the distance between the cells to the size of the foot, is such that the greatest distance the foot can slip is not over one inch.

Ice and Foot-hold.

Of course, when a pavement is completely covered and buried out of sight, with a layer of ice, it is not expected that its foot-hold, cells or grooves, will prevent the horse from slipping. The horse will slide on ice, of course, unless he be sharp shod, whatever be the character of the pavement that rests some inches below.

Foot-hold between Rail-tracks.

The Brooklyn city Railroad Company has laid a stone pavement between the rails on the up grade curve on Fulton street *expressly with reference to giving the horse the best possible foot-hold.* And the blocks selected, in size and shape, are more similar to the Nicolson blocks than those of any other pavement ; and the grooves extend in a straight line transversely from rail to rail. So that, except the difference of material and the rounded edges of the blocks, the surface and grooves are like the Nicolson.

(See suggestions relating to foot-hold on previous pages, from 38 to 44.)

Third. Advantage of the Nicolson Pavement, as relates to the general form of the Surface of the Street and Gutter.

So far as it is desirable to give the general surface any particular form, the Nicolson is *especially* adapted; for being composed of wood, its blocks and other parts can be worked into any required shape with the greatest facility.

This feature in regard to gutters is of no little importance.

As pavements are generally laid, the deepest point of the gutter lies close up to the curbstone, which in some respects is quite objectionable.

1st. The streets being oval and teams suddenly sheering up toward and then away from the curb, the vehicles are swung around against the unyielding curb with great violence, which often snaps axles, crushes wheels, and dashes mud upon the sidewalk.

2d. Vehicles standing close to the curb are often smashed by being run into; when if there were a few inches between them and the curb, the passing vehicles, though coming in contact, would jostle by without doing them any harm.

3d. The gutter being lower than the surface of the pavement, stages and heavy teams, getting into them, have, many times, much difficulty in sheering out of them again.

4th. The union between the curb and pavement being the lowest line of the street, the water washes out the dirt and causes the curb stone to lean towards the street and separate from the flagging.

5th. Though the sidewalk and curb stone are intended as a protection to pedestrians against horse and vehicle, yet when the lowest point of the gutter lies close to the curb the sidewalk is encroached upon by the greasy hubs of wheels, and often with a continuous pile of snow and mud thrown up from the pavement.

6th. The gutter being lower than the rest of the pavement, tips the tops of vehicles down and towards the sidewalk, which makes the getting in and out of close carriages quite inconvenient.

To avoid these objections the gutter should have its lowest point or line run *eight or nine* inches from, and then slant up to, the curb as shown in figure 6 (page 68) illustrating the gutter of the Nicolson.

Such a gutter as this, by means of the Nicolson, is made without additional expense; and prevents the vehicles from encroaching upon the sidewalks, and gradually eases up the side resistance of the curb against the slewing vehicle; thus saving many a wheel and axle, and avoiding danger of capsizing, and preventing the side chafing of wheels against the curb, and preventing vehicles from encroaching upon pedestrians by running too close to the walk.

Fourth. Advantage of the Nicolson Pavement, as relate to Material, or the peculiar Merits of Wood over Stone, for Pavements.

So far, in the discussion, attention has been called to *surface, foot-hold, forms, construction,* &c., and to the several advantages resulting from having these requisites properly provided; without especially referring to the material of which the pavement might be composed.

All must admit too that the requisites, as heretofore described, are desirable and indispensable; and that the Nicolson supplies all *these* qualities in a far superior manner to anything else. In fact, it is only by means of *wood* that even *these* requisites can be so perfectly obtained. Yet the *great and peculiar* advantages of the Nicolson over other pavements result, not alone from its construction and the supplying of the requisites as, so far, set forth, but from the *nature of the material* employed.

It would seem that the first conception of a pavement took the form of a cobble stone, and ineradicably impressed itself upon the mind.

Cobble stones were easily found; they could be had for the gathering and the freight, which, in times past, cost little or nothing. Coasters and shipmasters many times gathered them into their unladened vessels and brought them to the seaport cities as ballast. But vessels, as a general thing, having come to be freighted both ways of their passage, and cobble stones becoming scarce, and so more expensive, together with the fact that the Belgian blocks can be had so cheaply and so near to New York, have given the Belgian pavement quite an introduction, within a few years past. Nevertheless, this pavement, too, is composed of hard rough stones.

Cobble stones should never have been brought to the cities; but, being on hand, they should be employed to fill in with, for docks and holes, the only use for which they are suitable. If they are wholly unfit for a pavement in some streets, like Broadway and Fifth avenue, then too they are equally unfit for any street, as John and Pearl streets, for instance. Why not? For no vehicle enters or leaves Broadway without traversing other streets.

The idea that pavements must be composed of something very hard and resistant, like stone or iron, we think can be shown to be a fallacy—a popular error.

1st. Advantages of the Nicolson, or Wood over Stone, as relate to the Horse.

Pavements are for horses and vehicles. Though many vehicles (horse cars) run on iron tracks, yet all horses must go on the pavements. And vehicles, not being animate, are incapable of suffering and abuse; and if the pavements be rough, hard, slippery or what not, they can be made strong, light, heavy, &c., as may be required, to adapt them to such pavement. Not so with the *horse.* Man must take *him* as he is. Beyond shoeing him, he can do nothing. His structure and organism are unalterable, and therefore, any possible adaptation of horse and pavement to each other, must come by suiting the pavement to the horse.

In determining the material for pavements best suited to the horse, we have first to observe the natural adaptation which heaven itself arranged and made perfect. The horse in his wild state feeds upon grass which grows on soft earth (not rocks) and which, *with* the grass, makes a foundation as harmless and comfortable to the horse as a velvet carpet is to a man.

Yet humane and religious man, to gratify his selfishness, has not only captured and enslaved the noblest animal in the world, but has dragged him off of the great green carpet spread down by the good Architect for his wild prancing snorters, and compelled him, under the lashing abuse of heartless Jehus, all his life, to drag the burdens of men over the rough and stony street.

But we must have the horse ; we cannot dispense with him ; nor can we give him his native sod to travel on. But we can try to give him something better than solid, bare, rough rocks, which unavoidably jar his limbs, bruise his feet and ossify his joints. On these rock-like pavements he is not only cruelly abused, but half his usefulness is destroyed. So economy and humanity both call for a pavement better adapted to his nature, if it be *possible.*

And since horse and wheel *must* be iron shod, the paving material employed must be sufficiently hard and resistant to sustain the action of shoe and tire ; and yet be of such a nature as to modify the concussion—action and reaction—so as to take off that stunning, rock-like shock, so ruinous to bone and muscle.

Efforts have been made to save the horse from the injury of this severe concussion, by introducing rubber between his hoof and shoe ; but this is impracticable and too expensive.—To prevent jar and noise, rubber has also been introduced between felloes and tires, but without success.

We must at least give the horse a fibrous substance to plod his way upon, and so lessen the cruelty we *must* inflict.

Hence, as wood is the only feasible fibrous substance for the purpose, we must either employ it or stick to stone, or plunge back into the mud and dirt street.

Travelling on stone constantly is unnatural to any animal, especially to the horse, but it is not so much a violation to traverse wood.

The wood, set up endways of the grain, has the effect to modify the action, or the concussion, between the iron-shod hoof and itself, so as to make it quite different to the horse. The bruising and ossifying effect is not so great. Though the shoe makes no more perceptible impression on the wooden block than on the stone, yet there is a different result to the horse. A wooden pavement, even, is less injurious to the horse when he travels on the ends, than when the action comes on the sides, of the fibers ; yet the wood will resist and endure endwise of the fiber many times longer than sideways ; and, hence, the double superiority of the Nicolson pavement over a plank road.

As before explained, in the mechanic arts the relations of cause and effect are, sometimes, as unseen and unaccounted for, except in results, as in any other department of science.

All through the mechanic arts it is well known that the results of concussion are modified by bringing materials of an unlike nature together. A wooden maul and an iron hammer of the same weight may have the same *power* in driving a chisel to frame a building, or to drive a stone cutter's implements, yet the results of the forces are different. Sudden cooling of steel hardens it ; heating, softens it. Of two musical instruments, apparently exactly alike, one will be worth many times the other. A wagon made entirely of iron, though its weight and strength be the same as one made entirely of wood, will have a very different effect upon the passengers they carry. In fact, the effects of applying the same billet of wood, endwise or sidewise, to any purpose are as different as if materials of an unlike nature were employed ; both as relate to the results produced *by* and *upon* the wood ; whether the mode of action be that of concussion or friction.

And so it is with wooden paving blocks. Though not soft enough to be impressed, yet they have a marked modified effect on the wear and tear of horses.

There is no doubt but the same horse would endure, other things being the same, a *third* longer on the Nicolson than on any other pavement.

City horses are turned out in the country to recover from sore feet and the general bruising they have received on the stone pavements. A common saying is, "don't drive so fast on the pavement, or you will knock your horse to pieces:" another is, "hold him in until you get off the pavement:" another, "hold him up now you've struck the pavement."

Why is it so important to have such good tracks on race courses and parks, where horses but seldom go, and *never* for purposes of draft, and then have such rough stony tracks where countless horses must travel from morning to night, and from night to morning, and drag heavy burdens?

It is common sense—self-evident—that the horse, adapted by nature to roam on the soft green sod, is incapable of travelling all his life with impunity on an unyielding rough stony surface.

2d. Advantage of the Nicolson, or Wood over Stone, as relates to Noise, Rents, &c.

The noise of the city thoroughfare is one of its greatest drawbacks—its intolerable, yet immovable, nuisance.

And though the busy tread and constant roll of wheels over the paved streets of the metropolis will necessarily create at best a terrible clatter and racket; yet the phrase, "*thundering city*," so often used should be applied only to a cobble paved city; and the expression "*deafening city*," to a Belgian paved city;

while the term "*humming city,*" might be applied to a Nicolson paved city.

A completely noiseless pavement is not possible, and, as it would be unsafe, neither is it desirable; yet the thundering, deafening, rattling clatter of the stony street is not only disagreeable, but a great hindrance in the transaction of business. Along the thoroughfare it requires a trying, straining effort to converse. City stages, though mounted upon the best of compound springs, make such a rattling noise that passengers are as unable to converse in them as they would be in the midst of whirling spindles and flying shuttles of a clattering cotton mill.

The Fifth avenuers of our beautiful cities may erect their princely dwellings; adorn them with costly pictures and frescoes; and, with medalion and velvet carpets, render all movements about their apartments silent and sylph-like; and, to catch the cooling breeze of summer days and nights, fold back their lace and satin curtains, throw up their French plate windows, and gather about for social converse and pleasure; yes, on either side and from end to end of the thoroughfare, the inmates of every palace may thus seek relief and quiet; yet one coal cart alone rattling along the stony way interrupts all conversation, and destroys half the comfort and pleasure sought throughout the magnificent avenue.

In trafficing thoroughfares the inconvenience and interruption to business is immense. Counting-rooms and offices, which require the best ventilated and lighted parts of the building, are often located, to get rid of the clatter of the street, in some dark disagreeable, gas-lighted room.

An incontestible evidence that a relatively noiseless

pavement is required and appreciated for business purposes, is in the fact that, wherever the Nicolson pavement has been laid, the same rooms and stores have rented for a third more than they were before.

In San Francisco, where the Nicolson was laid in one of two adjoining parallel streets of equal standing in travel and traffic, and the Belgian at the same time laid in the other, the Nicolson pavement so much attracted the travel and traffic from the other street that the property holders, *at their own expense*, to save their rents, and trade, took up the Belgian and put down the Nicolson. This has been the result in other places, as well.

Many large real estate holders, instead of complaining of taxes, say they can afford to lay it every year, at their own cost, and then save money out of the increased rents, which the tenants are willing to pay in order to secure it.

In case of sickness the stone pavements are very objectionable; not only on account of noise, but the *jar* of the vehicles, which the stone communicates to the sensitive and diseased nerves of the patient. A rich person can sometimes influence permission, and afford the expense of having tan bark, or some other fibrous substance spread over the pavement around his dwelling; but the countless poor, when sick, must suffer and bear the racket and jar of the noisy rocky street. The major part of such suffering would be prevented by a general use of the Nicolson pavement.

There are many people, nominally in health, that cannot, for the jar and noise of stone pavements, reside in the city at all. While the constant irritation, produced by the ceaseless rumbling thunder of the city, more or less exhausts the nervous energy of all persons, even the most robust.

Yet the cost of one ordinary painting, or lace curtain, or carpet, or one pleasure wagon, or one horse, or one fluted column, or fancy door-cap of a store or dwelling, or the extra annual rent of a suit of rooms, or the pleasure of transacting business one year in a quiet store, or the comfort of relative quiet during one siege of sickness would be ample to lay the Nicolson pavement in front of any store or dwelling every year.

3d. Advantage of the Nicolson, or Wood over Stone, as relates to Iron Shoes and Tires.

As the use of iron shoes and tires is absolutely indispensible, the materials selected for pavements must have reference to this fact and the nature of iron. For the action, or wear and tear of the pavement on these is just as much a matter of consideration as *their* action, or wear and tear on the pavement.

As this matter of wear and tear of shoes and tires, however, will be better understood after alluding to the desirability of the Nicolson, it will be omitted here.

4th. Advantages of the Nicolson, or Wood over Stone, as relates to the Shaping and Laying of its Parts.

The several parts of the Nicolson, being composed of wood and bounded by planes and right angles, are brought into the desired shape with greater rapidity and perfection than it were possible, if the material was hard and the forms irregular. The implement employed for this purpose being, probably, the cheapest, most perfect and effective, in its results, of any labor-saving machine yet invented—which is the buzz-saw. One man, with one of these powerful saws, will do as much execution, in forming wooden paving blocks, as a regiment of men will do in getting out stone blocks.

Besides, the perfection of similarity in their shape and size, renders the labor expended in laying the pavement equally simple and cheap. For the most unskilled labor is competent to lay the blocks with accuracy and perfection; which enables nearly all the cost of the pavement to apply to the procurement of the most suitable material; instead of expending the labor in *forming* and *laying* a material, which, in itself costs but a trifle, and which, when laid, fails to answer any of the requirements of a good pavement.

5th. Advantage of the Nicolson, or Wood over Stone, as relates to Water and Gas Pipes.

The multiplicity of underground gas and water pipes, in cities of the present day, requires a pavement which can be easily taken up in small sections, and securely replaced.

The Nicolson, being impervious to water and moisture, keeps the earth dry beneath; and, being a good non-conductor of heat, the frost never reaches below it. So, wherever it is taken up, the ground is in as good condition, to be handled and repacked, in the winter as in the summer.

The right angle character and uniformity of shape and size of the blocks, together with the plank substructure, enable the sections taken up to be replaced with a perfect fit, and on a level with the general surface.

To prevent it from settling, a lap-joint plank is run under the edge of the undisturbed pavement and allowed to project under the edge of the returned section.

To remove a section of this pavement, first take up the blocks, and then, by means of a hatchet, cut off the planks. The same planks can be returned, by using a lap-joint plank, as above explained.

6th. Advantage of the Nicolson, or Wood over Stone, as relates to its Sanitary Properties.

The Cobble or Belgian pavements allow free infiltration of water and decomposed vegetable matter, which, thus escaping the evaporative power of sun and air, still undergo other changes and exhale in the form of poisonous miasmata.

But, by means of the geometrical symmetry of its parts, and their complete coating of asphaltum and coal tar, the Nicolson is perfectly impervious to water and moisture, and thus prevents infiltration of decomposed vegetable matter, and obviates all miasmatic exhalations.

Besides, by the use of the asphaltum and coal tar the streets are provided, throughout their breadth and length, with a lasting and valuable disinfectant.

This superadded sanitary quality of the Nicolson pavement, in large cities, is of great value, and, in these days of precautionary science, should not be overlooked.

From the many high scientific testimonials of the sanitary merits of this pavement, received by its inventor, we select the following:

BOSTON, November 27, 1866.

SAMUEL NICOLSON, Esq.:

DEAR SIR—I am familiar with your Patent Wood and Pitch Pavement, and with most of the improvements you have made n it, and can confidently express an opinion on its adaptation to the cities of our Southern States, and to those of the West India Islands.

It not only does away with muddy streets and makes a clean and comfortable roadway, but it also obstructs the rising of pestilential emanations from decomposing vegetable matter in the soil, but also chemically, if coal tar is employed in the composition, the phenic acid of coal tar being one of the best known antisceptics, meeting and decomposing any malaria that may come from below.

This opinion, which I originally expressed to you in my letter of December 5, 1854, has been, as I learn, amply verified by subsequent experience in Chicago, Toledo, and other of our Western cities.

If so simple a device as a good pavement will render healthful, cities which are now annoyed by pestilential effluvia from the soil, your invention will prove one of the most efficacious sanitary applications that has yet been made, and I am strongly impressed with the belief that such will prove to be the result in many cities of the Southern States, and in the West and East Indies.

I have seen your pavement in the streets of San Francisco, California, and know that it is highly approved there, and that an imperious demand for its extension throughout the city existed while I was there last year; for all agreed that it was the best pavement in the city.

Respectfully your obedient servant,

CHARLES T. JACKSON, M. D.,

State Assayer to the Commonwealth of Massachusetts.

No. 20 STATE STREET, BOSTON,
November 28, 1866.

SAMUEL NICOLSON, Esq.:

DEAR SIR—I have had abundant opportunities for observing the excellent qualities of your pavement while subjected to *constant* use, and its special adaptation to situations where the road bed is naturally moist

Apart from other features of this invention, and deserving of attention, is the *sanitary influence* of this construction.

In the cement used by you the powerful disinfectant carbonic acid is found, and the vapors of this substance will neutralize miasmata and destroy the germs of mould vegetation, which are constantly exhaling from the surface. We may confidently expect to find the vicinity of the pavement protected from those diseases produced by miasmatic exhalations, and in this view the substitution of your pavement for other kinds becomes a great public benefit.

Respectfully,

A. A. HAYES, M. D.,

Cons. Chemist and State Assayer.

7th. Advantage of the Nicolson, or Wood over Stone, as relates to the Pleasure of Riding.

If such magnificent and costly roads, as those of Central Park for instance, can be afforded and considered necessary, merely for the pleasure of riding and driving ; why should not the *pleasure* of riding and driving constitute a still greater consideration in business thoroughfares and resident avenues; over which there *must* pass a *hundred* vehicles, horses and riders to *one* that traverses the roads of Central Park. No vehicle can even reach these same pleasure roads of the Park, without first passing, many of them, for *miles* over the rough and jolting pavements of other streets and avenues. Why arrange a public garden so as to compel all its visitors to travel over a bed of thorns to reach a bed of roses? Why lay a rough rocky pavement in Fifth avenue, one of the long, narrow, crowded, yet magnificent gateways to this pleasure-driving Park?

Again: pleasure is not so much promoted separate *from*, as in connection *with* business. A man who drives and rides eight or nine hours a day, with heavy loads, over ruts, holes and gullies, down town, cannot make up for the wear and tear and discomfort, by hitching up to a light wagon and taking an hour's turn in the Park—can't *rough* it over the Belgian and Cobble stones *six* days in a week, and then heal the bruises and smooth the wrinkles with a little expensive "pleasure" driving Sunday—might as well say, transact business, and *suffer*, in your counting-room all day, in mid-winter, without any fire, to save the expense of stove and coal, with a view of making up for the suffering by sitting around a good fire at home in the evening.

The Nicolson pavement is no less appreciated by the hackman, stage-driver and teamster, as an *every-day* comfort, than the Central Park roads are by sportsmen and opulent pleasure-seekers, as a *now-and-then* recreation.

The agreeableness of riding on smooth roads is made up principally by the steady view it affords of the surrounding scenery; for, whenever we ride, we are in the midst of an actual live picture; a constant view of which we cannot avoid. If this great panoramic picture is made to appear as if it were being constantly jerked up and down, by being ourselves jolted over a rough road, half the pleasure of beholding it is, of course, destroyed. Hence, in this respect, the motion of riding is agreeable just in proportion as the track approaches a perfect plane.

Therefore it is, together with the fact that wood diminishes the noise, softens and modifies the concussion between itself and the shoe and tire, that the Nicolson affords that peculiarly agreeable, swinging, floating motion, almost as perfectly as the floating of an open boat in still water, which intensifies the pleasure of riding to the utmost degree.

8th. Advantage of the Nicolson, or Wood over Stone, as relates to its Universality.

The Nicolson is suited equally well for all kinds of roads. It is alike adapted to the heaviest vehicles and loads, or to the lightest pleasure buggies; to the crowded business streets of the city, or its boulevards and park-roads; for private walks and bridle roads; or for country thoroughfares, as a substitute for plank and McAdamized roads.—In short, wherever it can be afforded, it is good for any vehicle; even from the boy's velocipede to the iron founder's ponderous truck; and for the foot of man and every beast, from a dog to an elephant.

9th. Advantage of the Nicolson, or Wood over Stone, as relates to Cost—Fitness of greater importance than Cheapness and Durability.

In this age and country, the saying, "penny wise and pound foolish," is too often made the rule of action, instead of that better axiom that "what is worth doing *at all*, is worth doing *well*."

This is the case when cost takes no account of durability, and vice versa; for, cost and durability being strictly relative considerations, a thing, in the abstract, might be very cheap, yet dear; or dear, yet cheap; depending upon the relative durability. And so, in the abstract, the durability of things depends upon the relative cost. And, as these must be considered with reference to each other, so must *both* of these be estimated with reference to still other qualities; such as feasibility, necessity, comfort, pleasure, destruction, &c.; in short, everything embraced in the idea of general fitness.

The truth of this statement is too evident to need illustration. Yet this idea of relative importance of different qualities has its limits. That is; a thing might be cheap, but good for *nothing* at all; or, being good *every* way, yet *so* dear, as not, possibly, to be afforded; and so on. For instance, a coat, made of sheet-iron, lined with zinc and wadded with saw-dust, might be at once cheap, durable and warm; yet, for want of suitableness, utterly unfeasible. So it would be with cast-iron boots and tin hats. A wooden rail-track would be cheaper and more comfortable than an iron one, yet absolutely impracticable for want of strength, as well as durability. Golden dishes, though more suitable and desirable than fragile porcelain, are, for general use, impossible, because of insufferable cost. Lead is good to resist acids, but it would make a poor stove; and though it would make a bad gun, it is excellent for bullets.

This adaptation of one thing to another, of means to end, causes to results, in *nature*, is special, universal and perfect. But in *art* it is otherwise; there being more or less imperfection of fitness in all things, pertaining to the mechanic arts especially. Therefore, in judging of the qualities of anything, as suited to a given purpose, it is necessary to keep in mind all the desired requisites and conditions. Yet, generally, there are *some* qualities and objects which, in importance, decidedly preponderate over others. For instance, the *leading* objects of windows are, (so to express it) to keep out cold and let in light, to keep in heat and let out vision; as well as to let in natural heat and not let out artificial heat. These leading objects are so well accomplished by panes of glass, that their fragile nature, though rendering them *easily broken*, is never urged against them. If pieces of sheet iron were used, though they would not break, and might cost less, they would be opaque; and as transparency is *indispensable* and toughness *not*, of course, glass is selected. Yet glass would be still more perfectly adapted if the quality of toughness in the iron were added to its transparency; but this is impossible. *Hence*, if glass should cost many times its present price, still it would be the cheapest known substance for the purpose.

The cost and durability of an article have relation also to its destructive effects on other articles coming into contact with it.

For instance, if glass so affected the light as to produce insufferably sore eyes, it still would be unfit for windows. Suppose of two boots, one of them would wear out a stocking every day, and produce a new "corn" every four weeks; and the other one would wear out a stocking every month and produce no corns, at all. Who would say that the one boot were not cheaper than the other, though it cost ten to one? A run-away horse, purchased to-day, at quarter price, may, to-morrow, destroy the value of a kind and gentle one, by dashing to pieces a valuable wagon. A cheap but insecure harness, by breaking, the first time used, might cost the value, not only of horse and carriage, but even life and limb. Suppose a new sort of coal, for fuel, could be introduced into market for one-fifth the ordinary price, and as valuable as any other, only that some peculiar acid in it would destroy a new stove, burning it, every fortnight. Who would contend that such coal would be cheap at any price?

In comparing two different articles produced for the same purpose, if capital can be commanded to meet the purchase, it is no argument against the one or the other, to say it *costs* more, unless it be shown that its extra practical worth is not equal to its extra cost. And in estimating such worth, all practical considerations must be taken into account; even including ease, comfort, convenience and appearance.

Otherwise it might as well be contended, for instance, that a live horse cost more than a dead one, or a sound one more than a lame one, a handsome one more than a bad looking one; or that a fine beaver coat costs more than a linsey woolsey one; or that a paper shirt costs less than a muslin one; or that fresh meat is dearer than tainted; that a rail-car costs more than a wagon; a railroad, more than a turnpike.

Though these ideas are not new, and everybody acts upon them daily, in all the common affairs of life, yet they are often grossly violated; not only by individuals, in a small way, but by governments, on a grand scale; and, among others, this subject of pavements stands out as a remarkable instance.

Within certain limits, the cost and durability of pavements are of minor consideration, compared with the many other far more important requisites.

Before cost and durability, come *flatness of surface*, and *foot-hold;* and above all, *fitness* as relates to the *organism of the horse;* then as relates to *noise, ease, comfort*, &c.

In fact, the general cost of pavements bears so small a relation to the many indispensable advantages of having streets well paved, that it seems simply absurd to say, in argument, that an acknowledged good pavement costs one or even *two* dollars, per square yard, more than an admitted poor one; or that the poor one will last a year longer.

But if cost and durability *can* be seriously urged as the leading considerations, then a mud street is best of all, for it costs less and lasts longer than any other. Yet it is not at all suitable, for reasons too obvious to be mentioned. Again, solid blocks of polished steel, though they formed a perfect plane, cost immense sums (or cost nothing) and last forever, would not be suitable, for quite as obvious, though altogether different, reasons.

If cost and durability of a pavement have any relation to its destructive effect on the objects that come into contact with it, then there is no limit to the reason why the Nicolson, or wood instead of stone, should be employed. For the wear and tear of shoes, horses and vehicles are many times greater on the cobble, Belgian and Russ, than on the Nicolson—on stone than on wood.—The wear and tear on the objects that traverse a pavement, under any circumstances, are a thousand fold greater than to the pavement itself. In other words, many thousands of dollars worth of shoes, horses and vehicles can be worn out on a few thousand dollars worth of pavement. Therefore the destructive effect of the pavement upon the objects moving over it, is more to be avoided than the destructive effect of horse and vehicle to the pavement.

Hence, any pavement that greatly increases the de struction of shoe, horse, and vehicle—ease, comfort and convenience is not economical, though it cost nothing and last forever.

It will be contended that tax payers will not consent to be taxed for a good pavement. This is presuming that the tax payer is "penny wise and pound foolish." For the Nicolson will save him seventy-five per cent. of his horse shoeing, fifty per cent. of the wear and tear of his horses and vehicles ; and increase the value of his real estate twenty per cent. ; to say nothing of the pleasure and comfort of having his property surrounded with pleasure roads, equal to those of Central Park.

The trouble is not in getting the tax payers to consent to be taxed, but in getting the city government to consent to the property holders laying the pavement, even at their own individual expense, that they may reap the benefits and profits of it.

Taking New York city, for instance, the relative cost of the Cobble, Belgian and Nicolson, is respectively $2, $3, and $4, per square yard.—Of course, where stone is dearer and wood is cheaper, there would be a double difference in favor of wood.

Besides, is there any argument why the Cobble should be abandoned, and one dollar a yard more paid for the Belgian to take its place, that is not equally applicable to show that the Belgian should give way for the Nicolson at an equal extra cost, if (all things considered,) it can be shown to be as much better than the Belgian as the Belgian is better than the Cobble? Why not say the Belgian costs a dollar per yard more than the Cobble, as well as to say that the Nicolson costs that much more than the Belgian.

If the tax payers prefer the Belgian at a dollar more, then they will prefer the Nicolson at a dollar more also; provided it be *worth* as much more than the Belgian, as the Belgian is worth more than the Cobble; of which there is not a shadow of doubt.

As "figures don't lie," and as some minds yield to no other argument—and not even to *this* unless they express dollars and cents—we here make an application of a little arithmetic.

Suppose a street to be 40 feet wide, and a building 20 feet wide; and we have for that building, (to pave its half of the street,) forty-four and forty-four hundredths square yards. Suppose the Cobble costs $2, the Belgian $3 and the Nicolson $4 per square yard; and that the several pavements will last ten years. This will give $44.44, difference between the Belgian and Nicolson for ten years, or $4.44 for one year.

There are, even scientific men, those who are rated as civil engineers and political economists, who, while they admit the superiority of the Nicolson, seriously contend, in print and public places, that the intelligent tax-payers will not submit to this paltry additional cost for a better pavement; which is presuming too much upon the stupidity of those who pay the bills.

Without again enumerating the many advantages of this pavement to the tax-payers, we assert with that confidence which is based upon actual figures, that any tax-payer, keeping a horse, (if there were no other pavement in his city,) would save money enough in a year, to meet this extra tax of $4 44, in the *shoeing of one foot of one horse.*

It is foolish to suppose, for a moment, that the intelligent tax payer, after freely contributing to build fine cities and introducing all the modern improvements of convenience and ornamentation, will now refuse to bear a small tax for this last and completing, this indispensable improvement. If he will bear a tax of two and three dollars for the Cobble and Belgian pavement, which ruin his horses and vehicles, diminish his pleasure and comfort, and that too without developing the full value of his real estate ; much more will he willingly pay an *additional dollar*, to save his horse and vehicle, promote his pleasure, ease and convenience, and to increase the value of his property.

In view of its being less destructive to the objects traversing it, the Nicolson, even at many times its present cost, would be cheaper than any other pavement ; while, at its nominal cost, it is the cheapest and most economical pavement which it is possible to invent.

10th. Advantage of the Nicolson, or Wood over Stone as relates to repairs.

In estimating the relative cost of two articles for a given purpose not only must the relative *first* cost and the relative durability be considered, but the comparative cost of repairs must also be taken into the account. And in this comparison, the Nicolson is also superior to all the stone pavements ; for, its blocks resting on an unyielding substructure, and being immovably supported in all lateral directions, it needs no frequent repairing here and there in spots, as do the Cobble and Belgian pavements.

This extra cost of repairing the stone pavements is far greater than the extra first cost of the Nicolson. And, as expense of repairs is part of the actual cost of anything, the Nicolson really costs considerably less than either the Cobble or Belgian.

11th. Advantage of the Nicolson, or Wood over Stone as relates to durability—General suggestions.

Wherever the Nicolson has been lai[illegible] the hundreds and thousands who gather around to witness the process, and the hundreds of thousands who pass over it, invariably exclaim: "That is just the needed pavement, if it will only stand." Every body appreciates it as answering all requirements, save that of durability. In fact, its numerous other merits, over stone and other pavements, are so apparent, that the pavement needs only to be seen to be pronounced a perfect success—"provided it will prove sufficiently durable."

Such expressions of doubt, however, come from only those who have no practical knowledge of the subject. This doubt of its durability by the masses who so readily perceive its other good qualities, comes from the fact, that wood, being so unlike stone, it seems incredible at first sight, that it can be substituted for it—especially for such severe use as paving city streets. This inference is so conclusive with many, that no amount of evidence would satisfy them short of an actual test. Yet there is conclusive scientific evidence of its durability.

But to satisfy all, even the most incredulous, there is not wanting either kind of evidence; for the Nicolson, as will be shown, is no new experiment; it having been extensively used in many parts of the country for years.

As there are several ways in which a pavement may fail, so there are, of course, several required features of durability; for it matters not whether a failure comes from one deficiency or another, so long as failure is the result. For instance, a pavement that does not continue to afford foot-hold cannot be considered durable, though it retain all its other qualities; and so with each of other needed requisites.

1*st. Durability as relates to the evenness of the general surface.*

The importance of a flat and even surface, and the necessity of a substructure to support it, have been discussed elsewhere. Hence it is only necessary in this connection to show how admirably the Nicolson maintains its evenness of surface.

It is apparent to any observer that, however great be the weight of a freighted vehicle, it never has but four points of contact with the pavement at a time. And these points of contact are so small that each of them will not occupy but a small area of a single block. Even the heaviest wheel may rest on the mere corner of a single block. Now it is also evident that this pressure cannot be rapidly transmitted from one block to another without depressing them, unless they have a larger area on the ground than their own independent bearing will afford; which is not the case with the Belgian or Cobble pavement. But the Nicolson, having a plank substructure, spreads out or extends the base of each block in all directions; in the aggregate, equal to the area of an entire plank. So that each block, in its turn as the wheel passes over it, has as much bearing on the ground as if it were sixty-four times its own size—that is, supposing the plank to be *one* by *twelve* feet, and the blocks *three* by *nine* inches. It is, therefore, impossible for one block to settle lower than another, unless the plank be broken in two, which is not possible, while it rests on the ground.

Hence the durability of the Nicolson against a disturbance of its general surface, or the formation of holes and gullies, is unsurpassed by any modern structure, not even excepting the Broadway Russ.

This want of uniformity of surface is one of the present defects of the Russ, notwithstanding so much expense and care to have the blocks well laid.

2d. Durability as relates to the Lateral or Side Pressure.

As each block in its turn is submitted to the forcible action of the entire pressure of the wheels passing over it, with often a powerful side strain, from both wheel and horse, it becomes necessary that each block be not only *well* secured, but, for the integrity of the whole pavement, it must be *immovably* staid in all lateral directions, as well as from beneath. The construction of the Nicolson pavement admirably supplies each of its blocks with this indispensable support.

The exact uniformity of the shape and size of the several blocks, together with their parallelogramic form, enables them to support each other. In fact, they fit together so perfectly that, when laid and cemented with the asphaltum and tar, they constitute, as regards lateral stability, a solid unyielding mass. Besides, the base of each block, being flat and at right angles to its sides, and resting on a flat substructure, there is no tendency of the blocks to cant over, from any amount of pressure that fails on the pavement in perpendicular lines.

No matter how flat and smooth may be a paving block, or of what composed, or however suitable, if it cants over, for want of side support, it fails to produce a level surface, and, so, a perfect pavement.

Hence this feature of durability is as important as any other.

And in this respect, too, the stability of the Nicolson far surpasses that of any other small-block pavement ever produced—that is, any pavement not composed of large slabs, such as constitute the upper layer of the Appian way.

3d. *Durability as relates to the General Form of the Street.*

The substructure and the perfect union between the blocks, together with the asphaltum and coal-tar cement, prevents the dirt from working up between, or out from under, the pieces; and, so, obviates the displacement of dirt below the pavement.

The lines of blocks extending from curb to curb, and affording each other such responsive bearing and resistance, together with the oval form of the street, gives the pavement, in maintenance of its general form, the support of the arch.

An English report, speaking of the new pavement of Cheapside and Holborn, (two of the principal thoroughfares of London,) says, "it is based upon broken granite instead of loose earth which is constantly working through the interstices, and vitiating the solid bearing which the stones should possess. A further security against its working into holes is given by dressing each stone accurately to the same breadth, and into the same form, so that each tier of stones spans the street like a bridge. This is an improvement on the Roman system; they depended for the solidity of their construction on the size of their blocks, which were irregularly shaped, although carefully and firmly fitted."

In the Nicolson the plank substructure takes the place of the broken granite.

4th. Durability as relates to Foot-hold.

Foot-hold should continue until the pavement is otherwise worn out. Yet stone pavements, like the Russ and Belgian, soon become polished as smooth as glass, and yield no foot-hold besides what can be obtained in the seams of the blocks.

But the fibers of wood, becoming filled with sand and gravel, do not admit of a smooth polish, like stone or iron. So that, even irrespective of grooves, the Nicolson affords better foot-hold than other pavements.

The *durability* of foot-hold is one of the peculiar merits of this pavement, for while foot-hold, in other pavements, fails *first*, in this it never fails, before the structure otherwise yields. The cells are from *three to four* inches deep, and their filling, being composed of gravel, asphaltum and coal-tar, works out so as to leave their depth, at all times, just right for the corks of shoes; and never any deeper.

The grooves or cells, therefore, of this pavement cannot fail, until the body of the blocks are worn away or down to the "strip," a depth of *four* inches.

The difference of the *quality* of foot-hold, on a stone and on the Nicolson pavement, is illustrated by the difference between a *new* and *old* file.

While the difference between the *durability* of the Nicolson and other pavements, in respect to foot-hold, is the same as the difference between two files, one of which soon becomes smooth and worn out, and the other remains as sharp and cutting, and so as good as new, until the body of the file itself is consumed or worn away.

5th. Durability as relates to Decay.

It is well known that coal-tar makes the most protective and durable paint in the world, in places exposed to the elements; for, being perfectly impervious to water, it protects from decay produced by changes of moisture and dryness. And, having no affinity for oxygen, it does not waste away by oxygenization and consequent disintegration.

As every part of the Nicolson—substructure, blocks, strips and filling—is completely saturated with, and poured full of this antiseptic, protective and preservative substance, it is not possible for the wood to decay during the period that any pavement can, consistently, be expected to last.

The asphaltum and coal-tar not only coats each piece of wood for its own protection, but it fills up the seams between the several parts of the pavement, and cements them all together, so as to render the whole structure impervious to water from above and moisture from beneath.

6th. Durability as relates to abrasion of the immediate surface or face of the Blocks.

We now come to speak of that feature of durability of the Nicolson pavement, respecting which there is so much doubt in the minds of those who have no knowledge, either of the power of wood to resist the forces of concussion and friction, or of the practical results of the Nicolson where it has been tested for years:—that is, the durability of the Nicolson, or wood, to prevent itself from being materially hammered to pieces, on the surface, by *concussion*, or worn away by *friction.*

The nature of the action or contact, to which pavements are submitted, is that of both concussion and friction, from hoof and wheel.

Now, it is well understood by all, practically versed in the mechanic arts, that where these two modes of action are required, it is always an object, either to *increase* or *prevent* attrition and disintegration. For instance, a stone-cutter seeks to wear away the surface of stone, with a grooved hammer, by the force of *concussion;* and an ax-maker wears away his axes to an edge, on revolving stones, by the force *of friction;* while the same stone-cutter strives to prevent the abrasion between his maul and chisel produced by concussion; and the machinist tries to prevent attrition, between all revolving parts on their bearings produced by the action of friction.

Hence whether the object be to *increase* or *prevent* abrasion and attrition, the first step is to ascertain the effect of one substance or material upon another when acting upon each other by concussion and friction. For some materials work admirably well together to *produce*, while others work equally well to *prevent* abrasion and attrition. For instance, a stonecutter, since he *must* employ steel chisels, seeks such material to act upon them as will prevent the disintegrating effect of concussion; not only on the chisels, but on the driver itself; hence he uses a wooden maul instead of iron. So with the farmer, he employs a beetle or wooden sledge, instead of an iron hammer, to act upon iron wedges. The stone-cutter, too, uses hardened steel instead of soft iron for hammers to batter away the stone; the blacksmith employs steel, instead of wood hammers, to batter down rivets, bolts, &c., by the force of concussion. The ax-maker employs stone instead of wood to *grind away* his instruments by friction. While the machinist makes gudgeons and their bearings of *unlike* materials to *diminish* the power of friction; and, for the same purpose, adds between them lubricating substances; as oil; while the stone-polisher, to *increase* the action of the friction, puts between the stone and rubber sand and water, instead of oil, and so on.

In selecting materials, however, that will work together to the best advantage to *promote* or *prevent* attrition and disintegration, it is generally the case, that the choice of substances is only on one side. That is, for instance, the stone-cutter must take the stone as he finds it, and adapt his tools to *it;* and, as steel chisels are *necessarily* employed, he seeks for such materials for a maul as are suited to the steel tools and itself. So he not only employs wood, but such wood, and uses it in such a manner (endways of the grain) as is the very best suited to protect the chisels and itself. So with the ax-maker, since he *must* employ steel for his axes, he selects not only stone, but such stone as will most effectually cut away the steel.

So, too, since hoof and wheel *must* be shod and bound with iron, there can be no choice of materials that shall act on pavements ; hence, such material, for pavements, must be selected as is best adapted to protect itself and the iron shoe and tire against the disintegrating effect of concussion and friction. And since more or less dirt and grit are always scattered over the pavement, and often water, (by rain and sprinkling,) these facts too must be borne in mind, in choosing a material best suited for a pavement.

Starting with the fact then, that iron shoes and tires are indispensable, and that the nature of their action on the pavement is that of both concussion and friction ; the next question to determine is, whether wood is not better adapted than either stone or iron to prevent attrition and abrasion, not only of the pavement itself, but of the shoes and tires, as well ; for the protection of these also, against extraordinary wear and tear, makes up the importance of this feature of durability.

If the Nicolson pavement had never been practically tested, and wood not employed in various other ways, to resist the disintegrating force of concussion and friction, it is true, it might naturally enough be inferred, that the durability of so soft, fibrous, and porous substance, for such purposes would, in *all* applications, be inferior to the harder and homogeneous substances of stone and iron.

But the Nicolson *has* been tested ; thoroughly and extensively ; and thus demonstrated to be a grand achievement, by the incontrovertible evidence of *success ;* to which must yield the most stubborn opposition and prejudice.

While there is abundant evidence in various

branches of the mechanic arts, irrespective of such practical test, to amply prove that wood will endure for pavements.

Besides; there are scientific principles with which to satisfactorily demonstrate the superior durability of wood.

1st. *The Evidence of durability drawn from a few of the many similar uses of wood, and other illustrations.*

All through the mechanic arts, where the forces of concussion and friction are brought to bear, if the object be to prevent attrition and disintegration—wear and tear—substances of dissimilar nature and hardness are brought together.

In case of *concussion*, for instance, the carpenter employs a wooden mallet to act on iron-shanked chisels, or iron hammers on wooden-shanked chisels; the caulker, a wooden iron-bound hammer to act on steel implements; the stone-cutter, a wooden maul on steel chisels; the farmer, a beetle or wooden sledge, on iron wedges, and so on.

Now if either of these drivers were to be made of a Belgian paving block, or any kind of stone, and submitted to the same use, it would not last one day; while some of these wooden drivers endure for many years; even a life-time.

In case of *friction*, for instance, the machinist employs, instead of iron, brass bearings for iron gudgeons; while, in some instances, wood or leather is more durable than either; as in tobacco machines, where grit gets between the bearings; or as in the case of turbine water wheels; the lower gudgeon of which, it is found, can be supported longer with a piece of wood set up endways of the grain, than by the hardest steel, or any other substance; the clock-maker causes a brass wheel to act on a steel verge.

Again, where the object is to cut away, disintegrate, wear and destroy the material by concussion, substances of a similar resistant nature, though of unequal hardness, are brought together.

For instance, the stone-cutter acts upon the stone with a hard roughened hammer, and steel chisels; and the steel engraver cuts away his plate with still harder steel implements; and so on.

And again, when the object is to increase attrition by *friction*, hard and resistant materials, of dissimilar hardness are brought together.

For instance, the hardest steel implements are rapidly worn away and brought to an edge, by the use of grindstones, whetstones, emery paper and the like. A tempered file cuts an untempered one; while the lapidary cuts diamond only with diamond; the stone-worker cuts away and polishes his work by an iron plate working on sand and water; large blocks of stone and marble are sawed by simple smooth blades of iron, sliding over sand and water; and so on.

These are but few of the illustrations of which there are thousands.

Now if these various materials were accidentally brought together without reference to their adaptation to produce desired results, we might put oil on the marble slab and on the grindstone; and make gudgeon boxes of granite, and lubricate them with sand and water; or make a farmer's beetle and carpenter's mallet of stone; or use wood for whetstones and grindstones; and the like.

How long time, for instance, would it require to sharpen a dull chisel, axe or scythe on a wooden grindstone; or to wear out a wooden grindstone, by an attempt to sharpen such implements on it; or how long time would it take to saw (endwise of the grain) a six foot cube of wood into boards, with a stone-cutter's gang of saws, compared with the time needed to saw, into slabs, a similar block of marble.

Even a stone-cutter's grooved or ten-ax hammer cannot cut away soft wood, endwise of the grain, as rapidly as it can the hardest granite; while the hammer, used thus, would last forever.

Notice now the similarity between some of these operations and the action of shoe and tire on pavements.

For instance the corks of shoes, and tires, together with grit and water, act on the stone pavement much like the stone-cutter's compound ax-hammer and polisher, and the sand and water.

The use of the hammer is to break off the inequalities of the surface, to prepare it for the polisher and sand and water; while the object of the whole process is, to smoothen and polish the surface.

To prove that the action of shoe and tire is similar; take a large block—say four feet square—of all kinds of stone, from soft free-stone to the hardest granite, just as they come from the quarry, and lay them level with the pavement in Broadway, and in a short time, each of them will become as *smooth as glass*.

This is the state of the entire pavement of Broadway to-day, as also of all other stone pavements, after being a short time laid—even the surface of the Cobble stones themselves become very smooth and slippery.

In this comparison, the *materials* and the *modes of action* are the same. The material in both cases acted *upon* is stone. The hammer is made of blunt steel blades, which answer to the *steel corks* of the shoe; the tire is steel or iron, and answers to the *polisher or rubber;* the grit and dirt of the street, sprinkle water and rain, answer to the *sand and water.* The force of the hammer is that of *concussion;* while the polishing process involves the action of *friction*; the action of corks and tires is that of both concussion and friction.

The polishing is done by cutting away the surface; and in the act or process of cutting away, the destruction or cutting is generally reciprocal. For instance, if a wooden grindstone, (to use the phrase) will *not* wear away a steel scythe, *neither* will the steel scythe cut away the wooden grindstone; but a stone grindstone, (to use the expression) *will* cut away the steel scythe, and so *will* the steel scythe cut away, or wear out, the stone.

So it is with a stone pavement. *It* will wear the steel corks and iron tires; and they, by their resistant nature, will cut and wear away the stone pavement.

But if a wooden grindstone will *not* wear away the steel scythe, nor itself be worn away by it; then will *not* a wooden pavement wear away steel corks and iron tires, or be itself worn away by them.

Numerous other similar illustrations might be given to prove, that wood is more durable for some purposes than the hardest homogeneous substances.

2d. *Evidence of durability drawn from general principles.*

The cellular and fibrous character of wood gives it altogether different physical properties from homogeneous substances. By acting upon the side of a billet of wood, by concussion, or friction, or both, the fibers will be easily broken and separated. But if these forces are brought to bear on the ends of the fibers, the wood will manifest great resistance and endurance. Notice how much greater is the force required to cut, mar or bruise a bit of wood on the end of the grain than on the side.

The reason why a block of even pine wood will endure the concussion-force and action of the stone-cutter's ten-ax hammer longer than a block of granite, is because the particles of the fibers, though softer, being tougher are not disunited from each other by the action of the hammer; but are hammered closer together. When the granite is acted upon, its surface-particles, though hard, are disengaged—disintegrated—separated from one another, and driven off.

The wood being so much softer than the hammer, it (the hammer) is not affected at all by its action on the wood.

When the fibers become well hammered down closely together the wood is said to be "broomed"; and when thoroughly "broomed," its power to resist the force of concussion, surpasses the credulity of those not practically acquainted with the subject. A single wooden maul will, sometimes, last a stone-cutter for *thirty* years, and be used ten hours a day, all the while.

Wood is also equally durable against the action of friction. The reason why wooden bearings (endwise of the grain,) will last longer than steel, iron or stone, is essentially the same as the one just given.

Again, the steel or iron gudgeon of a turbine water-wheel, for instance, is one of the most difficult things in machinery to find a durable bearing for. Every thing has been tried ; and wood, (endwise of the grain), is at last found to be the *most* durable.

The iron, in the friction-action, being harder, smoothens the wood ; and the wood being too soft to cut the iron, polishes it. This action goes on until the attrition is practically nothing. This principle is recognized in machinery by selecting materials of unlike hardness for gudgeons and their boxes or bearing, generally.

This principle of bringing substances of dissimilar nature and hardness together to prevent attrition, disintegration, wear, &c., has been recognized and acted upon from time immemorial, all through the mechanic arts.

And there is no instance where it is, or can be employed more scientifically, and therefore to better advantage, than in the adoption of wood for city pavements.

The exposed surface of the Nicolson is at first covered with a layer of coal-tar, sand and gravel.

By the time this is scattered, washed, and ground off, the surface of the wood seems to appear to the eye, and to be exposed to the actual contact of horse and vehicle upon its fibers ; but upon closer examination it will be found that the sharpest and hardest instrument will not penetrate, or even reach the grain or surface of the wood itself. The sand and gravel, adhering to the wood by means of the coal-tar, become partially pulverized, and driven into the grain of the wood, by the iron shoes and tires, in such a manner that the exposed surfaces are actually coated with an impenetrable layer of pulverized sand, which, though it completely protects the wood from wear, does not destroy its quality of preventing a severe concussion with the iron shoe and tire.

It is this coating of the wood with a mineral surface, combined with the brooming of the wood by the shoe and tire, that renders it impossible to drive a nail into the pavement after it has been a short time in use.

The same is true of a wooden beetle, &c. ; yet a nail can be driven into the same wood, endwise of the grain, easier than in any other way, before it is broomed.

The shoe and tire do not wear off this hard mineral surface, for the reason that its lower side is supported only by the grain of the wood, which does not place it between two hard substances, but it penetrates and follows down into the grain of the wood.

But, with the stone pavement, the iron and stone coming in contact by concussion and rubbing, and both being perfectly resistant in their nature, they necessarily pulverize and triturate each other.

Since steel edged hammers, iron rubbers, sand and water, are brought to act upon each other by the forces of concussion and friction, expressly for the *purpose* of disintegrating, cutting and wearing away the stone, it is absurd to make use of the same materials and forces—corks, tires, dirt and water, concussion and friction—where the first object is to *prevent* disintegration, cutting and wearing away—not only of the pavement, but of the corks and tires, as well.

If steel chisel and harthstone, steel scythe and grindstone, will cut and wear each other away ; then will steel corks and stone pavements, iron tires and paving stones, do the same. Steel and stone, iron and stone, reciprocally cut and wear—mutually destroy.

The laws of forces, of wear and tear, are not suspended to suit convenience, to humor prejudice, or satisfy selfishness. In these, as in all other matters, like causes produce like results.

3d. Evidence of durability from the durability of the ancient Roman Pavements.

Though the upper stratum of the ancient Roman pavements was flat and even and firmly supported, yet the *durability* of these structures was not owing so much even to these qualities of a good pavement as to the fact, *that this very principle*, of causing materials of an unlike resistant nature to act upon each other, was practiced. For, in those ages, as horse shoes and wagon tires were not known, organic and fibrous substances—*hoof* of the horse and wood of the wheel—were brought to act upon the stone pavement ; instead of iron shoes and tires upon *wood* pavement.

The action of iron-wheeled Broadway stages, New York trucks and the like, with their iron-footed teams, is far more destructive to stone pavements than would be unshod horses and wooden-wheeled chariots.

4th. Evidence of durability from the actual use of the Pavement.

The Nicolson has been in use over ten years. Wherever it has been employed it has given universal satisfaction. In those cities where it has been in use for several years, the rapidity or increase of its introduction and demand is as the square of the time. Where it has been used longest, the demand for it is greatest—the more it is had, the more it is called for. Which fact—unless it be supposed that the people of our most thriving cities are at work with their eyes shut, or are incapable of knowing what best meets their wants—is the strongest evidence, in proof of the superiority of the Nicolson, that can be adduced.

The following are some of the cities employing this pavement, viz: Chicago and St. Louis; Milwaukee, and other cities in Wisconsin; Detroit and Kalamazoo, in Michigan; Cleveland, Cincinnati and Toledo, in Ohio; Fort Wayne, in Indiana; Memphis and Nashville, in Tennessee; San Francisco and Sacramento, in California; Portland, in Oregon; and Elmira, in New York. It is now, also, beginning to be introduced into New York city, Brooklyn and Philadelphia, besides many other cities.

It was first laid in Chicago, in 1856; in St. Louis in 1860. The amount put down in St. Louis from January to October, in 1866, was about 60,000 square yards.

The amount ordered and contracted for in Toledo, for the year 1866, as stated by the Contractor, was *six miles*, or about 125,000 square yards; and so on.

A detailed account, of its introduction, amount laid, quantity ordered, satisfaction it has given, &c., in all the several cities is not necessary. But as this part of the evidence—the practical testimony—of its durability and success, will be principally relied upon by many, a somewhat detailed account of its introduction and success in one city, at least, may be desired. We select Chicago. The history and success of the pavement in this city, is well described in the following communication, which speaks for itself:—

CHICAGO, ILL., November 9, 1866.

Hon. GUY R. PELTON, *President Board of Commissioners for Improvememt of Broadway, New York:*

SIR—We have been requested by Hon. Charles E. Jenkins, of New York, to address a communication to you setting forth the chief facts within our knowledge concerning the Nicolson pavement in Chicago.

We annex hereto a tabular statement showing the exact number of square yards of this pavement which have been laid in Chicago each year since its first introduction.

This statement is made up from the sworn statements of the city officers and it is entirely accurate. We also annex a statement showing the amount of said pavement for the laying of which orders have been passed the City Council and which are still to be executed.

It will be seen, by referring to these tables, that the pavement was first introduced into Chicago in the year 1856, when a half block of it was laid on Wells street and the remaining half of same block in the year following—the whole block con taining some 1,700 square yards.

In the years 1858, 1859 and 1860 about 30,000 square yards were laid, including Clark street, one of our most travelled streets.

In 1861, 1862 and 1863 about 24,000 square yards were laid, including Lake street.

In 1864 and 1865 about 42,000 square yards were laid, including West Lake street; and the order for paving Wabash avenue passed and the work on it commenced.

In 1866 over 140,000 square yards were laid, including several of our streets of heaviest travel, such as South, Water, Canal and Wells streets, and in the same year, 1866, orders were passed by the City Council for the laying of about one hundred and eighty-six thousand square yards (being more than six miles).

The total number of square yards laid amounts to over 244,000 square yards, or something over eight miles of paving.

Nothing more conclusive than this statement could be adduced in evidence of the favor with which this pavement has been and still continues to be regarded in Chicago.

The fact that, after ten years of experience of it, over 140,000 square yards or nearly five miles of it should have been laid during this year and one hundred and eighty-six thousand square yards ordered to be laid, is the best possible testimony in regard to the estimation in which it is held by the public and city authorities.

It may be truly said that so far as our business streets and leading thoroughfares are concerned it has supplanted entirely every other kind of pavement in the public regard.

This result has taken place in spite of very great practical defects in the execution of a good deal of the work. A considerable part of the pavement has been laid by contractors who have taken contracts at a price far below what the work could be properly executed for, and have been disposed to make as poor a job of work as could by any means be got through inspection.

But the inherent advantages of the pavement over all others which have ever been laid here, and the confidence in its durability created by the experience had of the wear of that laid on Wells and Clark streets (which we will allude to more particularly hereafter) has caused it to maintain its position in the public estimation, and while the defective work of the contractors is strongly denounced the Nicolson method of pavement is still universally approved by our property holders and business men. This approval of it is specially decided and unanimous among our teamsters and drivers and the proprietors of our livery stables, on account of the special benefits which result to them from its use, as there is no pavement so favorable to horses and carriages as the Nicolson.

No difficulty is experienced in the rapid and easy removal and relaying and the repair of the pavement when required to be taken up for the purpose of laying or repairing gas or water pipes, and with proper care the pavement so relaid can be made as perfect as when originally laid.

In regard to the Durability of the pavement our experience here is derived mainly from the pavement on Wells street, Clark street and Lake street.

That which was laid on the north half of the block on

Wells street in 1856 (the first ever laid in Chicago) was taken up in July of the present year (1866) when that street was raised to grade, and relaid with the same kind of pavement. It was then in tolerable condition, and would probably have lasted for two years longer, making in all twelve years.

The travel on that street has been very heavy. The pavement was in close proximity to the Wells street bridge, connecting the North and South Divisions of the city at that point.

Wells street leads to the Galena Railroad, over which there is an immense traffic, and teams very heavily loaded with flour, grain, pork and merchandise passed over the pavement in great numbers. Owing to the level character of our streets it is the custom here to draw very heavy loads. Twenty-five barrels of flour is a very ordinary load for a pair of horses, making, with the team a weight of over three tons, and much heavier loads than this are often carried.

During business hours the travel over this pavement was very continuous as well as heavy.

Mr. Reuben Cleveland, who was formerly the Superintendent of Public Works for Chicago, has recently testified under oath that, in June, 1859, some blocks of this pavement on Wells street were taken up by direction of the city, under his supervision, for the purpose of determining in regard to the *Durability* of the pavement, with a view to deciding whether or not to construct more on the same plan. It was found on that examination that the blocks had worn down three-sixteenths (3-16) of an inch per year, and that they were perfectly sound in every respect. The same gentleman, in Nov. 1865, visited the same pavement and took out of it two of the blocks and one of the short strips. These, he testified, were then in as sound and perfect a condition as when first laid down, and had worn down one and one-half inches in the nine years, being a little more than one-eighth of an inch per annum. One of these blocks and the strip were recently sent by us to New

York, by the hands of Mr. Jenkins, and will by him be submitted to your examination.

The pavement on Clark street was laid in 1858. This street is one of our principal thoroughfares, and since the pavement was laid upon it has taken the greater portion of the heavy traffic in the southern portion of the city. Teams going to the Southern Michigan, St. Louis, and Pittsburgh Railroads pass over portions of it. The block between Randolph and Washington streets, near the public buildings, containing the city and county offices and the court rooms, is the theater of an incessant and heavy travel. This street was raised to grade (just previous to being paved) with about three feet of soft filling material, and, in the haste to get the immediate use of the street, the pavement was completed as fast as the street was filled, and before the filling had any time to settle The consequence was that a settling took place afterwards, the effect of which was to cause the rows of blocks to become curved and swayed out of position, in some places to the extent of nearly two feet deviation from the right line of their original courses as laid. This fact was established by the statements under oath of L. L. Bond, Esq., for many years a member of the Committee on streets of the Common Council. Notwithstanding this extraordinary swaying of the courses of blocks the integrity of the pavement was not impaired, but after eight years of such wear as we have specified it is still a good pavement and will, with repairs in some portions, continue serviceable for several years longer. It has during this period received no repairs of any importance. In our judgment the action of the pavement, under such a trial, as this on Clark street has been subjected to, is very strong evidence of its durability.

We are informed by the Chief Engineer of the city that in cases where the *blocks* have to be relaid, the wooden foundation beneath will generally be found in good repair and amply sufficient as the support for a new superstructure of blocks. This is an important consideration with respect to the expediency of laying the pavement upon any street where (as on our Broadway) it would be subjected to a very continuous and heavy travel, and where the interruption of the travel for any considerable period would be a matter of serious public inconvenience and of heavy loss and damage to business interests—because the mere removal of the blocks and the relaying of new ones could be effected in so short a time as not to involve any long delay in the use of the street

The longest continuous piece of this pavement in Chicago is that which has just been completed on Wabash avenue. This is upwards of two miles in length and makes a magnificent drive, which is the admiration of the city. It has become a great thoroughfare and is the favorite resort of our citizens for driving, and every fine afternoon will be found thronged with the finest equipages in the city.

The advantages of this pavement over every other which has been laid here are, that while it affords an excellent foothold for animals which prevents them from slipping, it at the same time possesses a smoothness of surface which renders it an easy roadway for the drawing of heavily loaded teams. The effect of this pavement in preventing injury to the feet and limbs of horses is, as compared with any stone pavement, a very important advantage, not only in view of the prevention of needless suffering and injury to the animals, but as a matter of economy to their owners.

Another very important consideration, so far as our principal business streets are concerned, is the freedom from noise which it secures. The pavement is as noiseless as is consistent with safety and, in the summer time especially, this feature of it is found to be a very great advantage over any stone pavement.

It is also a paving which can be very easily and with little expense kept clean and free from dust, and in winter the non-conducting materials of which it is composed operate to keep the ground beneath from being affected by frost.

These are the principal facts within our knowledge respecting the introduction and use of the pavement in Chicago and the advantages which it has been found to possess.

GOODWIN & LARNED.

Statement *of the whole amount of Nicolson Pavement laid in the city of Chicago.*

Time when laid.	Streets on which laid.	No. of sq. yards laid.
1856	Wells street North, half of block between Lake and S. Water.........	800 00
1857	do South do do do do do	932 00
"	Washington street, from La Salle to Clark.................	1,707 00
		3,439 00
1858-59	Clark street, from Randolph to Polk................................	24,551 00
1559	do do do do Lake..............................	2,091 00
1860	Central Ave. do S.Water do do	1,728 00
1859-60	Alleys in Blocks 38, 16 and 19 in Orig. Town.........................	1,576 00
		29,946 00
1861	Lake street, from Wabash Avenue to Bridge	16,263 00
1862	do do do do Central avenue...................	3,340 06
"	S. Water, do Franklin street do Lake street.....................	3,135 33
"	Wells street bridge and approaches	375 00
1863	Intersection of Clark and Randpolph streets..........................	597 03
"	do do Madison streets..........	567 30
		24,267 72
1864	West Lake street, from Water to Holstead...........................	13,868 45
"	Iceland Queen Engine House.........................	192 88
1865	Wells street, from Madison to Van Buren.............................	10,256 67
"	Wolcott street, from Kinzie to Michigan............................	1,065 33
"	Michigan, from Cass to Clark........................... .. .	6,742 99
"	Dearborn, from Madison to Monroe...................................	2,267 55
"	South Water, from Wabash Avenue to Michigan...	1,704 37
"	do do Clark do Franklin...........................	5,755 05
		42,256 29
1866	Wabash avenue, from Randolph to 22d street.......................	67,311 00
"	West Randolph, do Holstead to West Water	16,766 00
"	do Madison, do do do Bridge..................................	13,206 67
"	Van Buren, do State do River....................................	11,130 43
"	Wells street, do Van Buren to Taylor...............................	11,514 00
"	do do Lake to S. Water..................................	2,349 33
"	Van Buren, do State to Michigan..................................	1,580 22
"	Canal, do Madison to Lake..	6,798 79
"	Washington, do State to Michigan avenue	4,109 08
"	Sherman, do Van Buren to Harrison street....................	2,642 02
"	Griswold, do do to Polk...	4,172 05
"	Approaches to Rush street Bridge.	1,082 00
		142,661

RECAPITULATION.

In 1856 and 1857....................	3,439 00
do 1858-9 and 60..................	29,946 00
do 1861-2 and 3.....................	24,267 72
do 1864 and 5.......................	42,256 29
do 1866..............................	142,661 59
	242,570 60 Total sq. yds. laid.

Statement of the amount of Nicolson Pavement which has been ordered to be laid by the City Council, and recommended by the Board of Public Works of the city of Chicago since September 1866, *not yet executed or put under contract:*

1866.	STREETS WHERE ORDERED.	ESTIMATED NUMBER OF SQ. YARDS.
	Milwaukee avenue, from North Des Plains to North Halstead	5,431
	Monroe street, from State to Michigan avenue	3,079
	Blue Island avenue, from Harrison to Twelfth street......	17,262
	Halstead street, from Harrison to railroad crossing.......	21,992
	Milwaukee avenue, from West Indiana to Elston road....	18,610
	Halstead street, from Lake to Harrison..............	16,000
	La Salle street, from Madison to Jackson..............	8,381
	North State street,from Michigan street to Chicago avenue	11,000
	North Well street, from River to Kinzie................	1,802
	Michigan avenue, from Twelfth street to Twenty-second street........	30,000
	Pine street, from Michigan to Chicago avenue	11,000
	Haddock place in Block 19..........................	514
	South Clarke street, from Polk to Taylor...............	4,529
	Franklin, from South Water to Madison................	9,600
	North Clarke, from Bridge to Chicago avenue	19,425
	Milwaukee avenue, from Kinzie to Indiana.............	5,431
	La Salle street, from Washington to Madison...........	2,069
		186,128

GOODWIN & LARNED.

In an interesting article entitled "Chicago," published in the "Atlantic Monthly" for March, 1867, the Nicolson pavement is thus alluded to:

"The principal streets are now paved with stone, or else with that *ne plus ultra* of comfort for horse and rider, for passer-by and ladies living near—the Nicolson pavement.

* * * * * * * *

Some parts of the mind are well cultivated there. Chicago is itself a college to all its inhabitants. When we see a boy reading in Roman history an account of the Appian Way, we say that he is improving his mind. The Nicolson pavemen has ten times more thought in it than the Appian Way; why is not an urchin improving his mind who stands, with his hands in his pockets, looking on while the workmen arrange the little blocks and pour in the odorous tar?

The New York *Evening Post*, of February 27, says:

"The enterprise heretofore exhibited by Chicago in the raising of the grades of her streets has been unexampled, and we see that the same spirit still prevails. The Board of Public Works have laid out an immense sum in the way of bringing to grade and paving with the Nicolson pavement a number of streets, the cost of which would be appalling to almost any other city. The list embraces considerable portions of sixteen different streets, amounting in all to five or six miles. No other pavement than the Nicolson—which has there had its most thorough trial—seems to be thought of."

The same paper, at another time says:

"We notice that the Chicago papers contain an announcement that during the next year (1867) the Board of Public Works propose to lay about eight miles of Nicolson pavement on the principal avenues; and we notice also that the City of Memphis, which is the only municipality that owns the patent right for its own limits, advertises for proposals for about six miles and a half of the Nicolson pavement. It is evident that experience confirms the good opinions of those who have used the pavement."

Official Opinion—Correspondence between the Mayors of Brooklyn and Chicago.

Although any amount of further confirmative testimony of the success of the Nicolson in Chicago can be given, yet the following recent correspondence, taken in connection with the foregoing report, and statements will be sufficient.

As we are speaking only of Chicago, (as before intimated,) we will give only the Chicago response :

[From " The Brooklyn Daily Union," February 6th, 1867.]

In consequence of the great interest excited by the proposed introduction of Nicolson pavement into this city, and with the view of ascertaining its worth from actual experience, his Honor Mayor Booth addressed the following queries to the official representatives of various cities, where the Nicolson pavement has been tested, and has received the replies hereto appended:

MAYOR'S OFFICE, CITY HALL,
BROOKLYN, Jan. 16th, 1867.

SIR—I have been informed that you have the Nicolson pavement now in use in your city.

You will do me a great favor if you will inform me upon the following points:

First. What in your opinion was the principal reason influencing the introduction of the Nicolson pavement in your city?

Second. In your opinion is the use of it in your city likely to be continued; and if so, why?

Third. Have you had it in use long enough to test its durability; and whether you have or not, what is your opinion with reference to this point?

Fourth. Can stone be easily procured in your locality for the laying of the Belgian or any other stone pavement, and what would be the relative cost of these different pavements in your city?

If you will have the kindness to give me, if possible, a very early answer to these inquiries, I shall not only esteem it a great favor but will be very happy to reciprocate at any time.

Very respectfully,

Your obedient servant,

SAMUEL BOOTH, *Mayor.*

REPLY FROM CHICAGO.

OFFICE OF THE BOARD OF PUBLIC WORKS, CHICAGO, Jan. 24, 1867.

HON. SAMUEL BOOTH, *Mayor of Brooklyn :*

DEAR SIR—The Mayor of this city, J. B. Rice, Esq., has requested me to give you all the information I can in regard to the Nicolson pavement. In answer to your first inquiry I will say, there was no particular reason inducing an experiment in said pavement, other than the fact that we were dissatisfied with the stone pavements which had been laid here from time to time. In answer to the second query I will say, after a test of about ten years, we are so well pleased with the pavement that we put down no other kind. We think it quite as economical as the stone pavement, much easier to keep in repair, easier to keep clean, less injurious to horses and vehicles, and at the same time comparatively noiseless. For these reasons I think it will be continued in use.

Third query. In regard to its durability, we have found by actual test that it will wear, when properly laid, full eight years—the floor remaining sound, upon which new blocks may be set at considerably less than first cost.

In regard to your fourth query I will say that, although we have pretty good stone of various kinds, we have none that gives entire satisfaction. The difference in the cost between stone and wooden pavement is about thirty-three per cent., the wooden pavement costing about that much more than stone. Should you desire further information I will at any time be glad to give it.

Respectfully yours,

J. K. THOMPSON,
Superintendent Public Works.

From the foregoing Report, statements and correspondence it will be seen that the amount of Nicolson pavement laid and ordered in Chicago *alone* is nearly *half a million* square yards ; which, however, is but a small portion of the aggregate amount laid in other cities.

After so long and extensive use of this pavement, in so many of our Western cities, with such entire satisfaction, to both authorities and tax-payers, that larger and larger quantities of it are from year to year being ordered and put down, it is useless and foolish to contend that its utility and durability are not established beyond all cavil and doubt, and that, too, by the incontrovertible argument of practical use and success.

Economy of the Nicolson Pavement.

As previously shown, the difference in the cost of the Belgian and Nicolson, in front of a dwelling or store, in New York city, is $4.44 per annum; which would only just about pay for *one tire* or *one set of shoes*, or for mending *one broken axle.* Yet, were the Nicolson in general use there would be, to offset this paltry sum of $4.44, items of saving, amounting in the aggregate, to more than the difference between paying ten times the entire cost of the Nicolson against having the Cobble or Belgian put down for nothing.

Though it is impossible to state what would be the exact amount of saving, yet it must be apparent to any practical observer that the Nicolson would, at least, save *seventy-five* per cent. of the cost of shoeing; *fifty* per cent. of the cost of tires, and *forty* per cent. of the breaks, wear and destruction of vehicles generally: and *fifty* per cent. of the wear, injury and, destruction of horses. It would increase rents and, so the value of real estate from *ten to fifteen* per cent. It would diminish the injury to freight, and lessen the damage to clothing, dry-goods and merchandise generally. It would greatly increase the ease, pleasure and comfort of riding, walking, and transacting of business. It would save *ninety* per cent. of the repairing of pavements. It would diminish disease and promote health; increase capacity of thoroughfares; enable vehicles to bear, and horses to draw heavier burdens. It would also save, at least, *thirty* per cent. of the expense of cleaning the streets.

To say nothing of damage to freight and clothing, of pleasure, ease and comfort, of repairs of pavements, of cruelty to animals, of increased capacity of vehicles, horses and thoroughfares, or of the increased value of rents, and the many other benefits—yet, when it is considered how immense is the aggregate cost of *horses*, *shoeing*, *vehicles* and *cleaning of streets;* and how infinitely greater is the entire cost of these than the entire cost of pavements; and how much greater is the percentage of wear and tear of these, on *any* pavement, than that of the pavement *itself;* and, too, when it is further considered that such large percentages of saving of these greater costs can be saved by an extra outlay of one or two dollars per square yard for an improved pavement; *then* to object to such improved pavement on the ground of its extra paltry cost is the most gross and absurd violation of the first principles—of the very axioms—of political economy that can be committed. So absurd is this objection that it can hardly be believed that any person, however ignorant, would venture to advance it. Yet there are those even in high position as civil officers and engineers, who, biased either by prejudice or pecuniary interest, or some other extraneous influence, have publicly and seriously argued and set forth this objection, saying the tax-payers will protest against having an acknowledged good pavement that will cost only *one* dollar more per square yard, though it save them *other* costs, equal to ten if not fifty dollars per square yard.

It don't take long for tax-payers to see the folly of trying to save money by putting three cents in one pocket and losing a dollar from another—by putting in the spigot and drawing the bung.

The saving on the shoeing of *one horse*, or on the wear of *one vehicle*, or on the deterioration of *one horse*, would more than pay the extra cost of the Nicolson for one lot. The saving on these for one year would meet the extra expense of this pavement for several years.

The general and special fitness, economy and humanity, ease, comfort, pleasure and convenience of the Nicolson pavement make up a combination of interests and inducements which must soon demand its general introduction throughout the leading business thoroughfares, resident avenues and pleasure drives of all principal cities.

It is too *late* to contend against this long-needed and satisfactory improvement in pavements;—as well argue against printing presses, railroads, telegraphs, sewing machines and the like. The Nicolson pavement is no longer an experiment; its utility is established; and they who oppose its introduction stand in the way of the public good. And the day is not distant when Samuel Nicolson will be publicly esteemed one of the first benefactors of his age.

www.ingramcontent.com/pod-product-compliance
Lightning Source LLC
LaVergne TN
LVHW021420110826
845150LV00007B/2009

9781425508982